Baltic Polymer Symposium 2019

Vilnius, Lithuania
18-20 September 2019

ISBN: 978-1-5108-9526-3

Printed from e-media with permission by:

Curran Associates, Inc.
57 Morehouse Lane
Red Hook, NY 12571

Some format issues inherent in the e-media version may also appear in this print version.

Printed with permission by Curran Associates, Inc. (2019)

For permission requests, please contact Vilnius University
at the address below.

Vilnius University
3 Universiteto St.
LT-01513 Vilnius
Lithuania

https://www.vu.lt/

Additional copies of this publication are available from:

Curran Associates, Inc.
57 Morehouse Lane
Red Hook, NY 12571 USA
Phone: 845-758-0400
Fax: 845-758-2633
Email: curran@proceedings.com
Web: www.proceedings.com

BALTIC POLYMER SYMPOSIUM 2019

PROGRAMME AND PROCEEDINGS

VILNIUS,
LITHUANIA,
SEPTEMBER 18-20, 2019

International Advisory Board

Chairman

Prof. Ričardas Makuška (Lithuania)

Members

Prof. Juozas Vidas Gražulevičius (Lithuania)

Prof. Martins Kalnins (Latvia)

Dr. Janis Zicans (Latvia)

Prof. Andres Öpik (Estonia)

Prof. Andres Krumme (Estonia)

Organizing Committee

Chairman

Prof. Ričardas Makuška

Members

Assoc. Prof. Tatjana Kochanė

Prof. Saulutė Budrienė

Assoc. Prof. Aušvydas Vareikis

Dr. Jūratė Jonikaitė-Švėgždienė

Dr. Alma Bočkuvienė

Dr. Tatjana Krivorotova

Dr. Vaidas Klimkevičius

Mrs. Irina Gerasimčik

Mrs. Sonata Gailiūnaitė

The support of *Baltic Polymer Symposium 2019* by the following organizations and companies is greatly acknowledged:

MAIN SPONSORS

Lietuvos mokslo taryba

RESEARCH COUNCIL OF LITHUANIA
Gedimino av. 3,
LT-01103 Vilnius, Lithuania
http://www.lmt.lt

ThermoFisher SCIENTIFIC

UAB „THERMOFISHER SCIENTIFIC BALTICS"
A. Graičiūno str. 8,
LT-02241 Vilnius, Lithuania

INFORMATIONAL SPONSOR

Go Vilnius

GO VILNIUS | VILNIUS CONVENTION BUREAU
Gynėju str. 14,
LT-01109 Vilnius, Lithuania
www.vilnius-convention.lt

SPONSORS

UAB „LINEA LIBERA“
Mokslininkų str. 12A,
LT-08412 Vilnius, Lithuania
http://www.linealibera.lt

UAB „LITNOBILES“
A.Kojelavičiaus str. 17,
LT-11101 Vilnius, Lithuania
http://www.litnobiles.lt

UAB „PLASTIKSĖ”
Stoties str. 63,
LT-21366 Vievis, Lithuania
http://plastikse.com

UAB “VILDOMA”
Granito str. 3,
LT-02241 Vilnius, Lithuania
http://www.vildoma.lt

UAB „NEO GROUP”
Industrijos str. 2,
Rimkai, LT-95346 Klaipeda district
http://www.neogroup.eu

BALTIC POLYMER SYMPOSIUM 2019 CONFERENCE AGENDA

September 18

15.00 – 19.00	**Registration and Reception** Lobby of the hotel "Artis Centrum Hotels", Totoriu str. 23
16.30 – 18.00	**Excursion to the Central Building of Vilnius University** Universiteto str. 3
19.00 – 21.00	**Welcome Party** Artis Centrum Hotels, **Restaurant "Adelia"**

September 19

Registration and opening. Theater Hall of Vilnius University (Universiteto str. 3)

8:30 – 9:00	**Registration**
9:00 – 9:15	**Opening Ceremony**
Session 1	**Theater Hall of Vilnius University** (Universiteto str. 3) **Chairman prof. R. Makuška**
	Invited Lectures
9.15 – 9.45	**Marta Giamberini** **SMART POLYMERIC MATERIALS FOR MICROENCAPSULATION 24** Department of Chemical Engineering, Universitat Rovira i Virgili, Tarragona, Catalonia, Spain
9.45 – 10.15	**Per Martin Claesson** **BIOINSPIRED ADHESION POLYMERS – WEAR RESISTANCE OF ADSORPTION LAYERS 25** KTH Royal Institute of Technology, School of Engineering Sciences in Chemistry, Biotechnology and Health, Department of Chemistry, Division of Surface and Corrosion Science, Stockholm, Sweden. RISE Research Institutes of Sweden, Division of Bioscience and Materials, Stockholm, Sweden.
10.30 – 11.00	**Coffee break** Artis Centrum Hotels, **Carmen Hall** (Totoriu str. 23)
Session 2	Artis Centrum Hotels, **Aida Hall** **Chairman prof. J.V. Gražulevičius**
	Oral presentations
11.00 – 11.15	K. Kubik[1], J. Paluch[2], J. Gabor[1], K. P. Jasik[3], A. Kwaśniewska[4], <u>A. S. Swinarew[1]</u>

	3D SCAFFOLDS IN RECONSTRUCTION OF THE LARYNX 32 [1]Institute of Material Science, Faculty of Computer Science and Material Science, University of Silesia in Katowice, Poland [2]Department and Clinic of Laryngology, School of Medicine in Katowice, Medical University of Silesia in Katowice, Poland [3]Department of Skin Structural Studies, School of Pharmacy with the Division of Laboratory Medicine in Sosnowiec, Medical University of Silesia in Katowice, Poland [4]Department of Radiology, Medical University of Silesia in Katowice, Hospital SPSK M, Katowice, Poland
11.15 – 11.30	S. Varnaitė-Žuravliova[1], V. Skurkytė-Papievienė[2], A. Abraitienė[2], A. Sankauskaitė[2], J. Baltušnikaitė-Guzaitienė[1] **IMPROVEMENT OF THERMOREGULATORY AND FRAGRANCE PROPERTIES OF TEXTILES USED FOR ORTHOPEDIC PURPOSES 33** [1]Department of Textiles Physical-Chemical Testing, Center for Physical Sciences and Technology, Kaunas, Lithuania [2]Department of Textile Technologies, Center for Physical Sciences and Technology, Kaunas, Lithuania
11.30 – 11.45	S. V. Kostjuk[1,2], M. I. Hulnik[1,2], I. V. Vasilenko[1] **AQUEOUS CATIONIC (CO)POLYMERIZATION: A GREEN ROUTE TOWARD SUSTAINABLE ELASTOMERS 34** [1]Research Institute for Physical Chemical Problems of the Belarusian State University, Minsk, Belarus [2]Department of Chemistry, Belarusian State University, Minsk, Belarus
11.45 – 12.00	A. Serra[1], F. Gamardella[1], V. Sabatini[1,2], X. Ramis[3], S. De la Flor[4] **NOVEL DUAL-CURED THERMOSETS OBTAINED BY CONTROLLED ISOCYANATE-EPOXY/THIOL CLICK REACTIONS 35** [1]Analytical & Organic Chemistry Department, Universitat Rovira i Virgili, Tarragona, Spain [2]Chemistry Department, Università degli Studi di Milano, Milano, Italy [3]Thermodynamics Lab., ETSEIB Universitat Politècnica de Catalunya, Barcelona, Spain [4]Department of Mechanical Engineering. Universitat Rovira i Virgili, Tarragona, Spain
12.00 – 12.15	A. I. Gostev[1], D. M. Krygina[2], E. V. Sivtsov[1], V. A. Ostrovskii[1] **RAFT POLYMERIZATION AS A TOOL FOR OBTAINING BIOMEDICAL MATRICES ON THE BASE OF N-VINYL-SUCCINIMIDE AND 5-VINYLTETRAZOLE 36** [1]Saint-Petersburg State Institute of Technology, Saint-Petersburg, Russia [2]Institute of Macromolecular Compounds of RAS, Saint-Petersburg, Russia
12.15 – 12.30	I. Dobryden[1], T. Tokarski[1], M. Cortes Ruiz[2], G. Li[1] and P. M. Claesson[1,3] **NANOSCALE MAPPING OF INTERPHASE WITH AFM: POLYMER BASED NANOCOMPOSITES 37** [1]Department of Surface and Corrosion Science, KTH Royal Institute of Technology, Stockholm, Sweden

	[2]Department of Chemical Engineering, Grove School of Engineering, New York, United States [3]RISE Research Institutes of Sweden, Division of Bioscience and Materials, Stockholm, Sweden
12.30 – 12.45	M. Netopilík **SEC OF POLYMER WITH COMPLEX DISTRIBUTION OF MOLECULAR WEIGHT AND BRANCH-POINTS 38** Institute of Macromolecular Chemistry, Prague, Czech Republic
12.45 – 13.00	T. Panova, A. Efimova, A. Berkovich, A. Efimov **GRAPHENE OXIDE BASED POLY(VINYL ALCOHOL) NANOCOMPOSITE FILMS: CONTROL OF MECHANICAL PROPERTIES 39** Polymer Department, Faculty of Chemistry, M.V. Lomonosov Moscow State University, Moscow, Russia
13.00 – 14.00	**Lunch** Artis Centrum Hotels, **Restaurant "La Traviata"**
Session 3	Artis Centrum Hotels, **Aida Hall** **Chairman prof. P.M. Claesson**
	Invited Lectures
14.00 – 14.30	**Ugis Cabulis** **RECYCLABLE AND RENEWABLE RESOURCES AS MUTUALLY COMPLEMENTARY RAW MATERIALS FOR THE PRODUCTION OF POLYURETHANE FOAMS 26** Latvian State Institute of Wood Chemistry, Riga, Latvia
14.30 – 15.00	**Illia Krasnou** **ELECTROSPUN NANOFIBROUS MATERIALS FOR ENERGY STORAGE AND HARVESTING 27** Department of Materials and Environmental Technology, Tallinn University of Technology, Tallinn, Estonia
	Oral presentations
15.00 – 15.15	S. Mačiulytė, A. Strakšys, S. Asadauskas **INFLUENCE OF ALIPHATIC MOIETIES IN DIISOCYANATES ON CHAIN EXTENSION KINETICS OF ADIPATE MACRODIOLS 40** Department of Chemical Engineering and Technologies, Center for Physical Sciences and Technology, Vilnius, Lithuania
15.15 – 15.30	F. Gamardella[1], F. Guerrero[1], S. De la Flor[2], X. Ramis[3], À. Serra[1] **CHARACTERIZATION OF THE FUNCTIONAL PROPERTIES OF A NEW CLASS OF VITRIMERS BASED ON POLY(THIOURETHANE) NETWORKS 41** [1]Department of Analytical and Organic Chemistry, Universitat Rovira i Virgili, Tarragona, Catalonia, Spain [2]Department of Mechanical Engineering, Universitat Rovira i Virgili, Tarragona, Catalonia, Spain

	[3]Thermodynamics Laboratory, ETSEIB Universitat Politècnica de Catalunya, Barcelona, Catalonia, Spain
15.45 – 16.00	A. Kadashchuk[1,2], A. Vakhnin[1], A. Zhugayevych[3], A. Köhler[4] **IMPACT OF FILM MORPHOLOGY ON ELECTRONIC STRUCTURE IN THE PROTOTYPICAL SEMICONDUCTING POLYMER POLY-3-HEXYLTHIOPHENE 42** [1]Institute of Physics, National Academy of Sciences of Ukraine, Kyiv, Ukraine [2]IMEC, Leuven, Belgium [3]Center for Electrochemical Energy Storage, Skolthech, Moscow, Russia [4]Bayreuth Institute of Macromolecular Research (BIMF), University of Bayreuth, Bayreuth, Germany
16.00 – 17.00	**Coffee break & Poster session I.** Artis Centrum Hotels, **Carmen Hall**
19.00	**Gala Dinner at Entertainment & Leisure Center "Belmontas"** (Belmonto str. 17, Vilnius)

September 20

Session 4	Artis Centrum Hotels, **Aida Hall** **Chairman prof. A. Krumme**
	Invited Lectures
9.00 – 9.30	**Jose Antonio Reina** **BIOMIMETIC MEMBRANES FOR SELECTIVE ION-TRANSPORT 28** Departament de Química Analítica i Química Orgànica, Universitat Rovira i Virgili, Tarragona, Catalonia, Spain
9.30 – 10.00	**Maik Feldmann** **CARBON COMPOSITES FOR LIGHTWEIGHT ENERGY STORAGE APPLICATIONS IN MOBILITY 29** Hexagon Purus GmbH, Kassel, Germany
	Oral presentations
10.00 – 10.15	A. G. Ayankojo, J. Reut, A. Öpik, V. Syritski **DEVELOPMENT OF A MOLECULARLY IMPRINTED POLYMER-BASED SENSOR FOR ELECTROCHEMICAL DETECTION OF MACROLIDE ANTIBIOTICS 43** Department of Materials and Environmental Technology, Tallinn University of Technology, Tallinn, Estonia
10.15 – 10.30	A. Kidakova, R. Boroznjak, J. Reut, A. Öpik, V. Syritski **MOLECULARLY IMPRINTED POLYMERS AS SYNTHETIC ANTIBODIES FOR NEUROTROPHIC FACTOR PROTEINS DETECTION 44** Department of Materials and Environmental Technology, Tallinn University of Technology, Tallinn, Estonia
10.30 – 11.30	**Coffee break and Poster session II.** Artis Centrum Hotels, **Carmen Hall**
Session 5	Artis Centrum Hotels, **Aida Hall** **Chairman prof. R. Merijs Meri**
	Invited Lectures
11.30 – 12.00	**Juozas Vidas Gražulevičius** **GLASS-FORMING DONOR-ACCEPTOR MOLECULAR MATERIALS FOR ELECTROLUMINESCENT AND OXYGEN SENSING APPLICATIONS 30** Department of Polymer Chemistry and Technology, Kaunas University of Technology, Kaunas, Lithuania
12.00 – 12.30	**Mikhail Smirnov** **ELECTROACTIVE POLYMER HYDROGELS: SYNTHESIS, STRUCTURE AND APPLICATIONS 31** Institute of Macromolecular Compounds RAS, Saint Petersburg, Russia

	Oral presentations
12.30 – 12.45	J. An[1], A. Dėdinaitė[1,2], F. M. Winnik [3,4,5], P. M. Claesson[1,2] **TRAPPED STATES AFFECT TEMPERATURE-DEPENDENT INTERFACIAL PROPERTIES OF PIPOZ-CONTAINING POLYMERS 45** [1]KTH Royal Institute of Technology, School of Engineering Sciences in Chemistry, Biotechnology and Health, Department of Chemistry, Surface and Corrosion Science, Stockholm, Sweden [2]RISE Research Institutes of Sweden, Materials and Surfaces, Stockholm, Sweden [3]Department of Chemistry and Faculty of Pharmacy, University of Montreal, Montreal, Canada [4]WPI International Center for Materials Nanoarchitectonics (MANA), National Institute for Materials Science, Tsukuba, Japan [5]Department of Chemistry and Faculty of Pharmacy, University of Helsinki, Helsinki Finland
12.45 – 13.00	C. V. Boyneburgk, P. Sälzer, J. Fuchs, H.-P. Heim **INFLUENCE OF THE ADHESION ON THE MECHANICAL CHARACTERISTICS OF SELF-REINFORCED POLYMER-VENEER-COMPOSITES 46** Institute of Material Engineering, Polymer Engineering, University of Kassel, Kassel, Germany
13.00 – 14.00	**Lunch** Artis Centrum Hotels, **Restaurant "La Traviata"**
Session 6	Artis Centrum Hotels, **Aida Hall** **Chairman prof. U. Cabulis**
	Oral presentations
14.00 – 14.15	M. N. Gorbunova, D. V. Eroshenko **NEW SILVER NANOCOMPOSITES WITH CYTOTOXIC ACTIVITY 47** Institute of Technical Chemistry, Ural Branch of Russian Academy of Sciences, Perm, Russia
14.15 – 14.30	O. V. Zaborova[1], M. N. Chernikova[2], P. N. Veremeeva[1] **NONDESTRUCTIVE ADSORPTION OF LIPOSOMES ON THE SURFACE OF LATEXES 48** [1]Department of Chemistry, Lomonosov Moscow State University, Moscow, Russia [2]D. I. Mendeleev University of Chemical Technology of Russia, Moscow, Russia
14.30 – 14.45	A. Efimova[1], G. Krivtsov[2], N. Melik-Nubarov[1], I. Grozdova[1], A. Yaroslavov[1] **NANOCONTAINERS BASED ON CHITOSAN AND ANIONIC LIPOSOMES 49** [1]Polymer Department, Faculty of Chemistry, M. V. Lomonosov Moscow State University, Moscow, Russia [2]Mechnikov Research Institute of Vaccines and Sera, Moscow, Russia

14.45 – 15.00	P. Franciszczak[1], A. Błędzki[1], E. Pleskot[1], M. Radwański[2], A. Kovalovs[3], G. Japins[3], K. Kalnins[3], A. Chate[3], R. Merijs-Meri[4], J. Zicans[4] **POLYPROPYLENE REINFORCED WITH SHORT PET FIBRES – MANUFACTURING ISSUES AND IMPACT BEHAVIOUR 50** [1]Institute of Materials Science, West Pomeranian University of Technology Szczecin, Szczecin, Poland [2]Ekotex, Namysłów, Poland [3]Institute of Materials and Structures, Riga Technical University, Riga, Latvia [4]Institute of Polymer Materials, Riga Technical University, Riga, Latvia
15.00 – 15.15	M. Varkale[1], I. Bochkov[1], R. Merijs Meri[1], J. Zicans[1], T. Ivanova[1], A. K. Bledzki[2] **POLYPROPYLENE AND ORGANOCLAY NANOCOMPOSITE MECHANICAL PROPERITIES 51** [1]Institute of Polymer Materials, Riga Technical University, Riga, Latvia [2]Institute of Materials Science, West Pomeranian University of Technology Szczecin, Poland
15.15 – 15.30	I. Vitkauskienė **CHALLENGES OF PLASTIC PACKAGING N/A** JSC Plastiksė, Vievis, Lithuania
15.30 – 16.30	**Closing session & Coffee break** Artis Centrum Hotels
16.30	**Excursion to Vilnius Old Town**

Poster session I

1.	G. Kručaitė[1], D. Tavgenienė[1], Z. Xie[2], X. Lin[2], B. Zhang[2], S. Grigalevičius[1] **POLYETHERS CONTAINING 4-(CARBAZOL-2-YL)-7-ARYLBENZO[C]-1,2,5-THIADIAZOLE CHROMOPHORES AS SOLUTION PROCESSED MATERIALS FOR HOLE TRANSPORTING LAYERS OF OLEDS 52** [1] Department of Polymer Chemistry and Technology, Kaunas University of Technology, Kaunas, Lithuania, gintare.krucaite@ktu.lt [2] State Key Laboratory of Polymer Physics and Chemistry, Changchun Institute of Applied Chemistry, Chinese Academy of Sciences, Changchun, China
2.	T. Kirila, A. Smirnova, A. Blokhin, A. Razina, A. Tenkovtsev, A. Filippov **INFLUENCE OF ARM STRUCTURE ON THE SELF-ORGANIZATION OF EIGHT-ARMS STAR-SHAPED POLY-2-ALKYL-2-OXAZOLINES IN AQUEOUS SOLUTIONS 53** Institute of Macromolecular Compounds of Russian Academy of Sciences
3.	R. S. Bernard, G. Sych, J. V. Gražulevičius **SYNTHESIS AND PROPERTIES OF TRIPHENYLETHYLENE DERIVATIVES CONTAINING CARBAZOLE AND ACRIDAN MOIETIES 54** Department of Polymer Chemistry and Technology, Kaunas University of Technology, Kaunas, Lithuania, ronit.bernard@ktu.edu
4.	D. Blazevicius[1], D. Tavgeniene[1], S. Grigalevicius[1], D. K. Dubey[2], M. Singh[2], S. Sahoo[2], J. H. Jou[2] **BICARBAZOLE-BASED POLYMERIC NETWORK AS A MIXED HOST FOR EFFICIENT SOLUTION-PROCESSED RED ORGANIC LIGHT EMITTING DIODES 55** [1] Department of Polymer Chemistry and Technology, Kaunas University of Technology, Radvilen plentas 19, LT 50254, Kaunas, Lithuania, dovydas.blazevicius@ktu.lt [2] Department of Materials Science and Engineering, National Tsing Hua University, Hsinchu 30013, Taiwan
5.	S. Nasiri, M. Cekaviciute, J. Simokaitiene, A. Petrauskaite, D. Volyniuk, V. Andruleviciene, O. Bezvikonnyi, J. V. Grazulevicius **TRI AND TETRAPHENYLETHENYL, SUBSTITUTED CARBAZOLE DERIVATIVES: SYNTHESIS, CHARACTERIZATION AND EXHIBITING AIEE AS EFFICIENT HOLE-TRANSPORTING OLED EMITTERS 56** Department of Polymer Chemistry and Technology, Kaunas University of Technology, Kaunas, Lithuania
6.	D. Volyniuk[1], K. Leitonas[1], J. Simokaitiene[1], E. Skuodis[1], M.D. Thiyagarajan[2], U.M. Balijapalli[2], M. Pathak[2], K. Sathiyanarayanan[2], P. Arsenyan[3], J.V. Grazulevicius[1]

Poster session II

SMART POLYMERIC MATERIALS FOR MICROENCAPSULATION

M. Giamberini

Department of Chemical Engineering, Universitat Rovira i Virgili, Tarragona Spain

Microcapsules are small particles which contain an active agent or core material surrounded by a coating or shell. The encapsulation of materials for protection and phase separation has evolved into a major interdisciplinary research focus.[1] The utility of microcapsules for efficient cargo storage and targeted release is of considerable importance in self-healing materials, nutrient preservation, agricultural applications, fragrance release, and drug delivery.[2] One of the most challenging tasks and the ultimate purpose of developing delivery systems is to modulate the release of encapsulated cargo substances. Strategies such as heat treatment, ionic strength, magnetic fields and light-induced morphology change have been used to alter the shell density and integrity, and then to influence capsule permeability.[3] As one of the most interesting parts of stimuli-responsive capsules, photo-stimuli responsive capsules are capable of affecting their micro-/nano-structures in the form of remote control triggered by external light e.g., sun light, without requirement of direct contact or interactions. Moreover, triggering the release of microcapsules by light has a number of advantages over other external stimuli: (i) photons do not contaminate the reaction systems and they have very low or negligible toxicity in contrast to chemicals; (ii) the excitation wavelength can be controlled through the design of the photo-responsive molecule, (iii) it is easy to control the time and/or local excitation.[4] The development of such highly light sensitive vesicles is of great importance, especially in the fields of surface sciences and environmental applications, where sometimes light would be the only available stimulus to drive the systems.

In this presentation, some examples of light-sensitive polymeric microcapsules will be described, that also show how the wavelength of the triggering light can be moved from the UV to the visible region, depending on the desired final application.

References

1. Zhang J. et al. One-Step Fabrication of Supramolecular Microcapsules from Microfluidic Droplets // Science – 2012.- Vol. 335 – P. 690-694
2. Esser-Kahn A.P., et al. Triggered Release from Polymer Capsules// Macromolecules – 2011- Vol. 44 – P. 5539-5553.
3. Abbaspourrad A. et al. Polymer microcapsules with programmable active release// J. Am. Chem. Soc. -2013- Vol. 135, No.20-P. 7744–7750
4. Asanuma H. et al. Synthesis of azobenzene-tethered DNA for reversible photo-regulation of DNA functions: hybridization and transcription// Nature protocols-2007- Vol.2, No.1- P. 203-212

BIOINSPIRED ADHESION POLYMERS - WEAR RESISTANCE OF ADSORPTION LAYERS

I. Dobryden[1], M. Steponavičiūtė[2], V. Klimkevičius[2], R. Makuška[2], A. Dėdinaitė[1,3], R. W. Corkery[1], X. Liu[3] and P. M. Claesson[1,4]

[1] *KTH Royal Institute of Technology, School of Engineering Sciences in Chemistry, Biotechnology and Health, Department of Chemistry, Division of Surface and Corrosion Science, Drottning Kristinas väg 51, SE-100 44 Stockholm, Sweden*
[2] *Institute of Chemistry, Vilnius University, Naugarduko 24, LT-03225 Vilnius, Lithuania*
[3] *School of Chemistry and Chemical Engineering, Shaanxi Normal University, Xi'an 710062, China*
[4] *RISE Research Institutes of Sweden, Division of Bioscience and Materials, SE-114 86 Stockholm, Sweden*

Mussel adhesive polymers owe their ability to strongly bind to a large variety of surfaces under water due to their high content of 3,4-dihydroxy-l-phenylalanine (DOPA) groups and high positive charge. In this work we use a set of statistical copolymers that contain short poly(ethylene oxide) side chains that are anchored to the surface in three different ways: by means of i) electrostatic forces, ii) catechol groups (as in DOPA), and iii) the combination of electrostatic forces and catechol groups. A nanotribological scanning probe method was utilized to evaluate the wear resistance of the adsorbed layers as a function of normal load. It was found that the combined measurement of surface topography and stiffness provided accurate assessment of the wear resistance of such thin layers. In particular, surface stiffness maps allowed to identify initiation of wear before a clear topographical wear scar was developed. Our data demonstrate that the molecular and abrasive wear resistance on silica surfaces depends on the anchoring mode and follows the order: catechol groups combined with electrostatic forces > catechol groups alone > electrostatic forces alone. The devised methodology should be generally applicable for evaluating wear resistance or "robustness" of thin adsorbed layers on a variety of surfaces.

RECYCLABLE AND RENEWABLE RESOURCES AS MUTUALLY COMPLEMENTARY RAW MATERIALS FOR THE PRODUCTION OF POLYURETHANE FOAMS

U. Cabulis, M. Kirpluks, A. Fridrihsone, A. Abolins, A. Ivdre

Latvian State Institute of Wood Chemistry, Riga, Latvia, cabulis@edi.lv

Polyurethanes (PUR) are one of the most required polymers due to their versatility. Rigid PUR foams are applied on a large scale mainly as thermal insulating materials. Due to dwindling petrochemical resources, the production of polymers, including PUR foams, from renewable or recyclable resources is being actively investigated. Also the environmental pressures demand cleaner industry which results in active development of "green" chemistry – using cleaner processes, less energy and renewable or recyclable materials. There are 3 principal sources of raw materials are investigated in this report: natural oils (rape seed oil (RO) and tall oil (TO)), lignin and residues of polyethylene terephtalate (PET).

Latvian State Institute of Wood Chemistry has a more than 40 years' experience in polyol synthesis from renewable and recyclable feedstock, and also knowledges about PUR system development with targeted characteristics. In present overview we analyzed different feedstock, different synthesis methods influence on polyol characteristics, such as viscosity, OH-value, acid value, compatibility with blowing agents, which are important for further PUR foams development and production

RO polyols were synthesized by transesterication and transamidation methods, also epoxidation method were used wit idea to obtain polyols with higher functionality. Despite the fact that natural oils are suitable raw material for polyol production for PUR industry, natural oils are food crops and thus are the first generation feedstock. TO and lignin as pulp production by-products qualifies as the second generation feedstock and have an advantage as their production does not compete with the production of food.

TO polyols were obtained by esterification and amidation of TO with triethanolamine (TEOA) and diethanolamine (DEOA), correspondingly. In these processes obtained polyols have big disadvantage – low functionality. Another option is unprotected TO fatty acids epoxidation by *in situ* generated peroxyacetic acid. After that, highly functional bio based polyol was synthesized by functionalizing epoxidized TO fatty acids with TEOA. Lignin containing polyols were obtained in reaction with propilene oxyde (PO) by oxypropylation method in high pressure reactor. Also bio/recycled polyols from PET and RO or TO were synthesized in 2 step processes.

Rigid PUR foams were obtained from synthesized polyols by hand mixing method and also scaled-up and foamed by PUR spraying machines. Obtained PUR foams were characterized from the point of view as potential thermal insulation material. Such important characteristics, as density, thermal conductivity, compression properties, and flammability were tested.

Total renewable and/or recyclate content in synthesized polyols reached up to 60%. The high bio/recycled content combined with excellent mechanical characteristics and low thermal conductivity of produced end product – rigid PUR foams, make these polyols suitable for preparation of stable polyol systems for long-term storage with low carbon footprint.

Acknowledgements. The financial support of the European Regional Development Fund Contract No 1.1.1.1/16/A/031 "Rigid Polyurethane/Polyisocyanurate Foam Thermal Insulation Material Reinforced with Nano/Micro Size Cellulose" is gratefully acknowledged.

ELECTROSPUN NANOFIBROUS MATERIALS FOR ENERGY STORAGE AND HARVESTING

I. Krasnou, T. Plamus, V. Vassiljeva, S. Malmberg, E. Tarasova, A. Krumme

Department of Materials and Environmental Technology, Tallinn University of Technology, Tallinn, Estonia, illia.krasnou@ttu.ee

Novadays, environmental pollution and natural resources depletion considered as the most serious reasons for development of novel and more effective technologies for energy harvesting, storage, transfer and utilization [1]. However, most of the innovative technologies and materials implemented in high-tech industries, what have indirect impact on global problems solution. Biomedical technologies, space and electric transport industries, communication technologies consume the majority of up to date developments.
Recently fibrous nanomaterials found application in electrochemical energy storage devices. They show excellent properties and performance: extremely large surface area and porosity, high length-to-diameter ratio, outstanding mechanical properties, and wide functionality. Electrospinning is a relatively cheap, simple, and versatile method for nanofibers production from diverse materials, organic and inorganic, natural and synthetic.

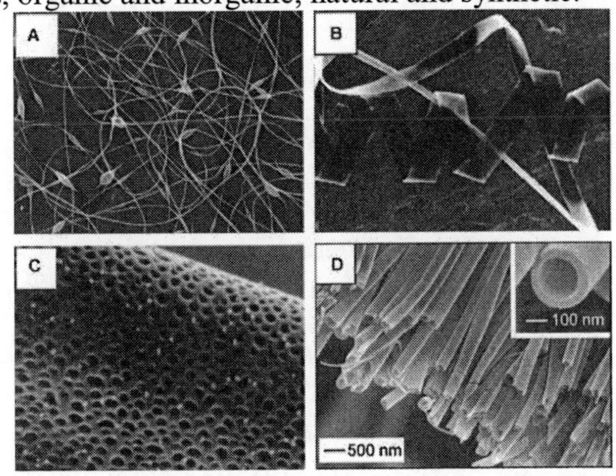

Fig. 1. Morphologies of electrospun fibrous materials [2].

Electrospinning as production method has significant advantages, such as:

- unique physical and chemical properties due to high surface-to-volume ratio
- continuous macroscopic material made of nanofibers
- bottom-up manufacturing method
- simple, versatile, and low cost producing method.

As well, it has challenges:

- theory of the electrospinning process is not enough developed
- low production rate
- strong demand on development of nontoxic solutions
- luck in uniformity in nanofibers below 50 nm diameter
- difficulties in device assembly from flexible nanofibrous materials

All these questions will be discussed at conference report.

Acknowledgements.
The European Space Agency, ESA contract number 4000119258/16/NL/CBi "Fully electrospun durable electrode and electrochemical double-layer capacitor for high frequency applications".

References

1. X. Shi, et al., Journal of Nanomaterials, **2015**, Article ID 140716 (2015).
2. Z. Dong, et al., Journal of Power Sources, **196**, 4886–4904, (2011).

BIOMIMETIC MEMBRANES FOR SELECTIVE ION-TRANSPORT

J. A. Reina

Departament de Química Analítica i Química Orgànica, Universitat Rovira i Virgili, Tarragona, Catalonia, Spain, joseantonio.reina@urv.cat

Substitution of fossil fuels has generated an increasing need of alternative energy sources: for this reason, the use of fuel cells has progressively gained importance, since they convert chemical energy directly into electrical energy. In proton exchange membrane fuel cells the electrolyte is constituted by a membrane, which is able to conduct protons. Among these devices, in the case of direct methanol fuel cells, methanol is used as a fuel, which exhibits several advantages over fuel cells based on hydrogen and compressed natural gas. Proton exchange membranes (PEM) constitute also a crucial component in water splitting devices, which can allow hydrogen production from "clean" energy sources, like in artificial photosynthesis. However, PEM with optimum characteristics (proton conductivity, water management, mechanical, thermal and oxidative stability, etc.) are far to be produced, and perfluorosulfonic acid membranes like Nafion® are still considered the reference material, despite their drawbacks. We have tackled this problem by preparing membranes based on self-assembling columnar side-chain liquid-crystalline polymers (SCLCPs) which lead to the formation of biomimetic ionic channels. That shows advantages for high performance in systems where the selectivity is needed.

Biological systems often show proton transport in a highly selective way, sometimes even in absence of water. Ion transport through cell membranes frequently uses proteic channels. Even though highly effective in cell membranes, they are inadequate for technological applications, due to difficult purification and complex chemical modification. One possible approach to obtain an ionic channel, could be based on forcing a polymer chain to adopt a helical conformation: here, we can find inspiration in Nature. For instance, Tobacco Mosaic Virus self-assembles upon mixing of its individual components, leading to a cylindrical structure in which the proteins organize in a helical array by the exo-recognition of the specific protein shape. According to this strategy, our approach was based on supramolecular chemistry, that is, we designed SCLCPs constituted by a polyether or a polyamine main chain, side functionalized with dendritic groups. According to our hypothesis, the process of exo-recognition should induce the formation of a columnar liquid crystalline structure, in which the inner part of the column is formed by a polyether or polyamine chain which acts as an ionic channel.

We succeeded in developing an easy and reproducible procedure, also at bigger scale, to prepare homeotropically oriented membranes, i.e. with the columns perpendicular to the membrane surface, which is expected to be the proper direction for optimum proton transport. The method is based upon a thermal treatment which involves heating of a polymer film, previously prepared by phase inversion process, above its clearing temperature and subsequent cooling to room temperature at a very slow rate, so that the columnar phase is allowed to slowly grow on a proper support (e.g. teflon). In this way, the aromatic moieties in the dendrons are producing the driving forces in the self-assembly process for the final materials to be homeotropically oriented, because of π-π stacking of aromatic moieties.

As far as membranes characterization is concerned, permeability tests were performed, by using hydrochloric acid as a feed phase and different alkaline cations chlorides as a stripping phase. The tests gave satisfactory proton permeability values, comparable to Nafion and depending on the cation in the stripping phase in the case of the prepared membranes based on polyethers. Moreover, the results suggested a cation antiport transport mechanism, and also pointed to selectivity to cations according to their size.

CARBON COMPOSITES FOR LIGHTWEIGHT ENERGY STORAGE APPLICATIONS IN MOBILITY

M. Feldmann

Hexagon Purus GmbH, Kassel, Germany, maik.feldmann@hexagongroup.com

The energy transformation and the slogan zero emission are in full speed and are becoming more and more apparent in everyday mobility. Battery electric vehicles (BEV) are offered by almost all OEMs but are known to have drawbacks in charging and range. In contrast, vehicles with fuel cells and electric drive (FCEV) can be refueled in a few minutes (3 to. 5min) and have ranges of over 500km. Both make the system also interesting for trucks, buses, trains and other vehicles, in which batteries have so far reached their performance limits. So far, there are only a few passenger car models and a limited number of hydrogen filling stations.
In addition to the fuel cell, the central element of this system is hydrogen storage. This is stored either at 350 or 700 bar in a pressure vessel. Pressure vessels made of carbon fibre composite material have proven to be particularly suitable for mobile applications, because they are particularly effective in storing hydrogen. This means that in relation to the weight of the container the stored amount of hydrogen is significantly higher than with metallic containers and this has a positive effect on the total vehicle mass. A so-called Type 4 pressure vessel weighs only approx. 25% as much as a Type 1 pressure vessel made of metal. In addition to the high-strength fibres, which are impregnated with a thermosetting resin and processed in the wet winding process, a liner made of thermoplastic is used as a gas barrier. Metallic components are used at the interface between the vessel and the valve. In addition to the materials, the winding pattern has a significant influence on the vessel performance and must be adapted to the stress, whereby not only the burst pressure but also cyclic leads and impact loads are in the spotlight.

GLASS-FORMING DONOR-ACCEPTOR MOLECULAR MATERIALS FOR ELECTROLUMINESCENT AND OXYGEN SENSING APPLICATIONS

J. V. Grazulevicius[1], G. Grybauskaite-Kaminskiene[1], N. Kukhta[1], R. Pashazadeh[1], E. Skuodis[1], A. Tomkeviciene[1], R. Keruckiene[1], J. Simokaitiene[1], M. Cekaviciute[1], Xiaofeng Tan, D. Volyniuk[1], K. Ivaniuk[2], P. Stakhira[2]

[1] Department of Polymer Chemistry and Technology, Kaunas University of Technology, Kaunas, Lithuania
[2] Lviv Polytechnic National University, Lviv, Ukraine

Glass-forming molecular materials containing donor and acceptor moieties recently synthesized at the laboratories of the presenting author will be reported.

Derivative of 3-(trifluoromethyl)benzonitrile and 3,3'-bicarbazole was found to exhibit both TADF and exciplex-forming properties [1]. Warm-white OLED based on this material showed external quantum efficiency (EQE) of ca. 20 %.

The derivative of acridan and dicyanobenzene was found to be efficient TADF-emitter exhibiting both thermally activated delayed fluorescence and aggregation induced emission enhancement. Green OLED fabricated using this emitter exhibited maximum current, power efficiency and EQE of 68 cd/m^2, 62 lm/W and 22.5 %, respectively [2].

A series of carbazole-quinoxaline-carbazole derivatives exhibiting TADF and mechanochromic luminescence properties were synthesized and studied. Green-blue to green-yellow TADF OLEDs fabricated by solution processing demonstrated EQE up to 10.9% and luminance of 16760 cd m^{-2} [3].

By utilization of the derivatives cyanophenyl and ditertbutylcarbazolyl substituted triphenylbenzene with the different substitution pattern as host and guest of the emissive layer, deep-blue OLED based on triplet-triplet annihilation with EQE of 14.1% were fabricated [4].

Derivatives of thianthrene and benzophenone exhibited room-temperature phosphorescence and demonstrated oxygen sensing ability [5]

Acknowledgment. This research was funded by the European Social Fund according to the activity 'Improvement of researchers' qualification by implementing world-class R&D projects' of Measure No. 09.3.3-LMT-K-712

References

1. G. Grybauskaite-Kaminskiene G et al., J. Mater. Chem. C, **6**, 1543-1550 (2018).
2. E. Skuodis et. al., Org. Electron., **63**, 29-40 (2018).
3. R. Pashazadeh et. Al., J. Phys. Chem. Lett., **9**, 1172- 1177 (2018).
4. N. Kukhta et al., J. Phys. Chem. Lett. **8**, 6199-6205 (2017).
5. A. Tomkeviciene et al, submitted to Sensor Actuat B-Chem.

ELECTROACTIVE POLYMER HYDROGELS: SYNTHESIS, STRUCTURE AND APPLICATIONS

M. A. Smirnov

Institute of Macromolecular Compounds RAS, Saint Petersburg, Russia, Smirnov_Michael@mail.ru

Flexible, biocompatible, lightweight and stimuli-responsible electroactive hydrogels (EHs) are prospective materials for potential application as sensors, actuators, electrode materials, controlled drug delivery systems and scaffolds for tissue engineering. These materials are usually constructed as the 3D polymer network cross-linked with covalent or non-covalent (coordination, ionic, guest-host or hydrogen) bonds. In latter case, the reversibility of crosslinks provides self-healing ability for the hydrogels, which is intensively studied during last decade. The electrical properties in such systems can be provided by carbon nanomaterials (nanotubes, graphene) or intrinsically conducting polymers (polypyrrole, polythiophene and others). The 3D network structure, which is responsible for mechanical properties, can be provided either by electrically active component with suitable crosslinker or (more often) by the additional crosslinkable polymer, for example, prepared by 3D radical polymerization of the acrylamide derivatives; physical or chemical cross-linking of the natural polymers such as chitosan, alginate, cellulose derivatives and others. Thus, EHs is the combination of two components: electroconducting structures and 3D polymer network. There are several strategies for preparation of such materials: in situ synthesis of electroactive component inside the previously prepared polymer network; synthesis of 3D cross-linked structure in the presence of electroconducting component and simultaneous formation of electroconducting molecules and cross-linked polymer network. These synthetic strategies, together with additional methods such as freeze drying, using of nanostructured templates, 3D printing and electrospinning, allows to prepare electroactive hydrogels with wide variation of structural parameters, such as crosslinking density, reversibility of cross-links, swelling ability, porosity (specific surface) and percolation threshold of electrically conducting phase. Structure of the hydrogel influences on its main properties: electrical conductivity, ion mobility, specific capacitance (for electrode materials) and mechanical properties, which are of great importance for application of these materials.

Acknowledgements: This work was supported by the Russian Foundation for the Basic Research (RFBR), grant 18-03-01167 A.

3D SCAFFOLDS IN RECONSTRUCTION OF THE LARYNX

K. Kubik[1], J. Paluch[2], J. Gabor[1], K. P. Jasik[3], A. Kwaśniewska[4], A. S. Swinarew[1*]

[1] *Institute of Material Science, Faculty of Computer Science and Material Science, University of Silesia in Katowice, Poland, andrzej.swinarew@us.edu.pl*
[2] *Department and Clinic of Laryngology, School of Medicine in Katowice, Medical University of Silesia in Katowice, Poland*
[3] *Department of Skin Structural Studies, School of Pharmacy with the Division of Laboratory Medicine in Sosnowiec, Medical University of Silesia in Katowice, Poland*
[4] *Department of Radiology, Medical University of Silesia in Katowice, Hospital SPSK M, Katowice, Poland*

Implantation artificial larynx is one way to avoid problems after resection due to: cancer benign tumors, injures. Mentioned problems are: nonphysiological way of breath (tracheotomy tube is necessary) voice communications deficit, swallowing problems. Nowadays there are many devices which can replace function of the larynx. They can made from variety materials and dived into outside – electrolarynx and inside – implants of the larynx. Efficient implant have to be completely biocompatibility and personalized for each patient. Artificial porous scaffold larynx covered with collagen and chondrocytes is a new solution in this way of treatment figure 1. This work is focused on novel polymer blend material that could be used as filament in 3D printers and extrusion feedstock. For now there are only some materials, like a polylactide (PLA), copolymer acrylonitrile butadiene styrene (ABS), high impact polystyrene (HIPS), nylon or commercial plastics like laywood and laybrick, that are generally used in 3D printers. Instead of it we propose medical, biocompatible PC as a matrix for novel material, however disadvantages of this polymer would make it difficult to use in pure form.

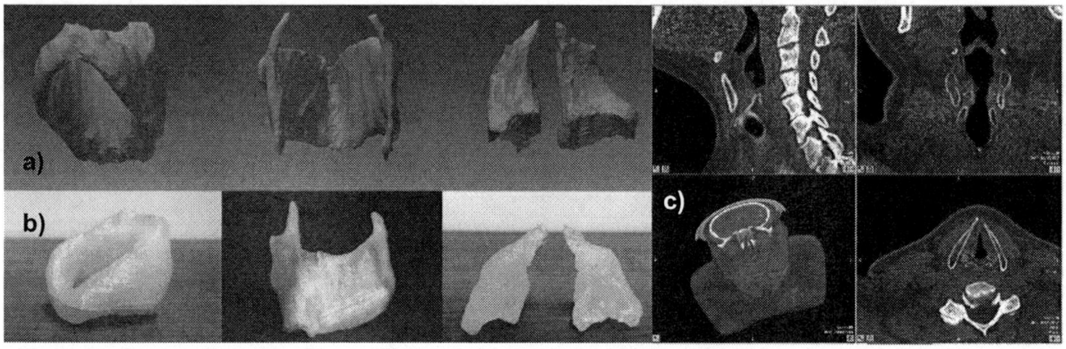

Fig. 1. Reconstruction of cartilage of larynx: a) microtomography model b) real model after 3D reconstruction c) tomography (CT) of neck in 3 layouts and volume scan.

References
1. G. Rijal, W. Li. 3D scaffolds in breast cancer research. Biomaterials 81(2016) 135-156.
2. D. G. Lott. Chapter 47 – Tissue Engineering of Larynx. Stem Cell Biology and Tissue Engineering in Dental Sciences. Academic Press: (2015) 625–640

IMPROVEMENT OF THERMOREGULATORY AND FRAGRANCE PROPERTIES OF TEXTILES USED FOR ORTHOPEDIC PURPOSES

S. Varnaitė-Žuravliova[1], V. Skurkytė-Papievienė[2], A. Abraitienė[2], A. Sankauskaitė[2], J. Baltušnikaitė-Guzaitienė[1]

[1] Department of Textiles Physical-Chemical Testing, Center for Physical Sciences and Technology, Kaunas, Lithuania, sandra.varnaite.zuravliova@ftmc.lt
[2]Department of Textile Technologies, Center for Physical Sciences and Technology, Kaunas, Lithuania

Thermal properties to textiles can be provided by integrating smart materials with phase changing materials (PCMs) in the form of microcapsules by various finishing techniques. PCM is able to self-absorb a certain amount of heat when the ambient temperature rises and give it back as the temperature decreases [1, 2]. Textile materials that absorb/emit odors improve the quality of textiles as well. Microcapsules of fragrances can be equated with a temporary container that is usually spherical and blocks liquid or gaseous fragrances, or can be compared to uneven structures if the fragrances are in solid form. Encapsulated flavours and other biologically active chemical compounds applied to textile materials give them cosmetic and comfort properties [2].
The aim of the study was to investigate the improvement of thermoregulatory properties and the refreshing smell of knitted fabric treated with different PCM microcapsules and to evaluate the longevity of these additionally granted properties.
In this work, a 17 % Polyester/83 % Cotton terry knitted fabric was treated with two kinds of PCM microcapsules in order to improve thermoregulatory and fragrance of orthopedic articles, i.e. splints, where knitted fabric was used in the inside of the article, surrounding human joints while wearing. Microcapsules MICRATHERMIC P (which improve thermoregulatory properties) were used separately and in conjunction with R-eSCENTial 310 Pure Menthol microcapsules (which provides menthol smell) to treat knitted fabric and to give an added value for splints. The effectiveness of microcapsules was investigated by differential scanning calorimetry (DSC) and by heating treated fabric (before and after 5 and 10 washing and drying cycles) with IR lamp. The thermal images of treated knitted fabric after irradiation with IR lamp were taken by FLIR InfraCAM. Fourier transform infrared spectroscopy (FTIR) and scanning electron microscopy (SEM) analyses were carried out in order to determine the longevity of microcapsules.
The results have showed that finishing of knitted fabrics with MICRATHERMIC P and R-eSCENTial 310 Pure Menthol microcapsules is long lasting, and remain effective even after 10 washing and drying cycles. These microcapsules can be used together, because they do not inhibit each other and do not affect each other's properties, and complement each other providing added value to the final product (see Fig. 1).

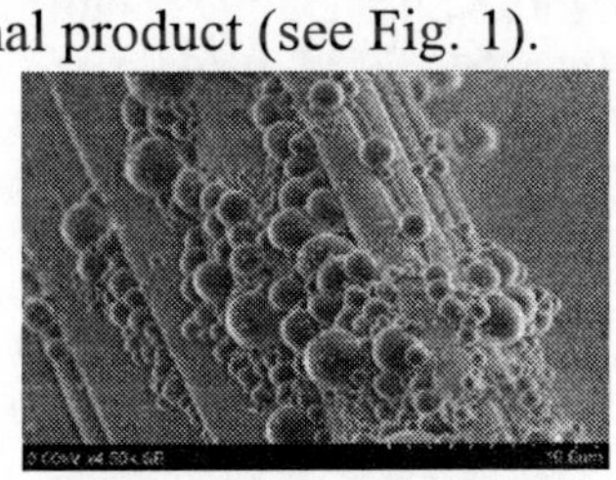

Fig. 1. SEM image of knitted fabric with MICRATHERMIC P and R-eSCENTial 310 Pure Menthol microcapsules, magnification x4500

References
1. M. K. Singh et al., Fibres Text. East Eur., **19(4),** 27-33 (2011).
2. S. Y. Cheng et al,. Res. J.Text. Apparel, **12(4),** 41-51 (2008).

AQUEOUS CATIONIC (CO)POLYMERIZATION: A GREEN ROUTE TOWARD SUSTAINABLE ELASTOMERS

S. V. Kostjuk[1,2]**, M. I. Hulnik**[1,2]**, I. V. Vasilenko**[1]

[1] *Research Institute for Physical Chemical Problems of the Belarusian State University, Minsk, Belarus,* kostjuks@bsu.by
[2] *Department of Chemistry, Belarusian State University, Minsk, Belarus*

Cationic polymerization in aqueous emulsion is believed to solve the main problems of conventional cationic polymerization, i.e. the necessity to work at low temperatures and under strictly anhydrous conditions.[1] This approach could be considered as a green alternative to the conventional cationic polymerization, which requires using the toxic chlorinated solvent.
In this work, the new catalysts consisted of complex of sodium dodecyl benzene sulphonate with rare earth metal, i. e. water-dispersible Lewis acid surfactant combined catalysts (LASC*s*) will be presented.[2,3] LASC initiates the emulsion cationic polymerization of isoprene affording high molecular weight polyisoprenes (M_n ~10^5 g mol^{-1}), predominantly with *trans*-1,4-microstructure.[2] The copolymerization of isoprene (IP) with styrene (St) results in statistical copolymers with monomodal MWD and single T_gs at St:IP molar ratios below 75:25. The copolymers synthesized at larger molar ratios, however, possessed two T_gs, the lower one for IP-rich blocks, the higher one ascribed to St-rich blocks. Such characteristics are the signature of thermoplastic elastomers.[2] LASC showed also high efficiency in the emulsion polymerization of β-myrcene, a naturally-occurring monomer. At 40°C, poly(β-myrcene)s of high molecular weight (M_n up to 150 kg mol^{-1}) and low glass transition temperature (T_g< –63 °C) were obtained. They possess predominantly 1,4-units (~43% *cis* and ~50% *trans*) and few 3,4-units (~7%) together with both regular head-to-tail (62 - 72%) and inverse (5 - 10% head-to-head and 23 - 28% tail-to-tail) sequences.[4] LASC showed also high efficiency in catalyzing the emulsion cationic copolymerization of β-myrcene and styrene, affording random copolymers with high molecular weights (M_n from 60 to 120 kg mol^{-1}) and single glass transition temperatures (T_g's from –43 to 15°C).[4] Finally, a simple procedure was developed to recover the catalyst from the reaction mixture, which showed high activity in the β-myrcene emulsion cationic polymerization even after five successive cycles of extraction.[4]

Acknowledgements. This work was supported by the State Belarusian Program of Scientific Research "Chemical technologies and materials" subprogram «Wood chemistry», project 4.1.12.

References

1. S.V. Kostjuk et.al., Acc. Chem. Res., **43**, 357–367 (2010)
2. I.V. Vasilenko et al., Angew. Chem. Int. Ed., **54**, 12728–12732 (2015)
3. I.V. Vasilenko et al., Macromolecules, **49**, 3264–3273 (2016)
4. M.I. Hulnik et al. Polym. Chem., **9**, 5690–5700 (2018)

NOVEL DUAL-CURED THERMOSETS OBTAINED BY CONTROLLED ISOCYANATE-EPOXY/THIOL CLICK REACTIONS

A. Serra[1], F. Gamardella[1], V. Sabatini[1,2], X. Ramis[3], S. De la Flor[4]

[1] *Analytical & Organic Chem. Dpt., Univ. Rovira i Virgili, C/ Marcel·lí Domingo, 43007, Tarragona, Spain, angels.serra@urv.cat*
[2] *Chemistry Dpt., Università degli Studi di Milano, Via Golgi 19 20133 Milano, Italy*
[3] *Thermodynamics Lab., ETSEIB Univ. Politècnica de Catalunya, Av. Diagonal, 08028, Barcelona, Spain*
[4] *Dept. of Mechanical Engineering. Univ. Rovira i Virgili. Av. Països Catalans 26, 43007 Tarragona, Spain*

In the communication, we report the preparation and characterization of a new family of thermosets obtained by thiol-isocyanate and thiol-epoxy reactions, both activated by temperature. The sequential dual character of this curing system relies on the faster reaction kinetic of the thiol-isocyanate coupling at lower temperature, than the thiol-epoxy reaction. The extent of the first curing step is controlled by the isocyanate/thiol equivalent ratio. This proportion defines the intermediate and final materials properties. Both reactions have a click character since they are specific, without the formation of undesired by-products, and they lead to homogeneous networks, which are very adequate to be used as shape memory smart materials, because of their narrow transitions that allows quick movements and a rapid change in their properties [1,2]. The formulations studied were formed by different proportions of diglycidylether of Bisphenol A and one of the two isocyanates, hexamethylene diisocyanate and isophoronediisocynate. Pentaerythritol tetrakis (3-mercaptopropionate), which acts as crosslinking unit of both type of network structures, was added to the formulation in stoichiometric ratio. As catalyst, 1-methylimidazole was used. The kinetics of both curing stages and the conversion achieved were studied by differential scanning calorimetry (DSC) and Fourier transform infrared spectroscopy (FTIR). Intermediate materials were characterized by calorimetry and fully cured samples by DSC, thermal dynamo mechanical (DMA) and thermogravimetric (TGA) analyses.

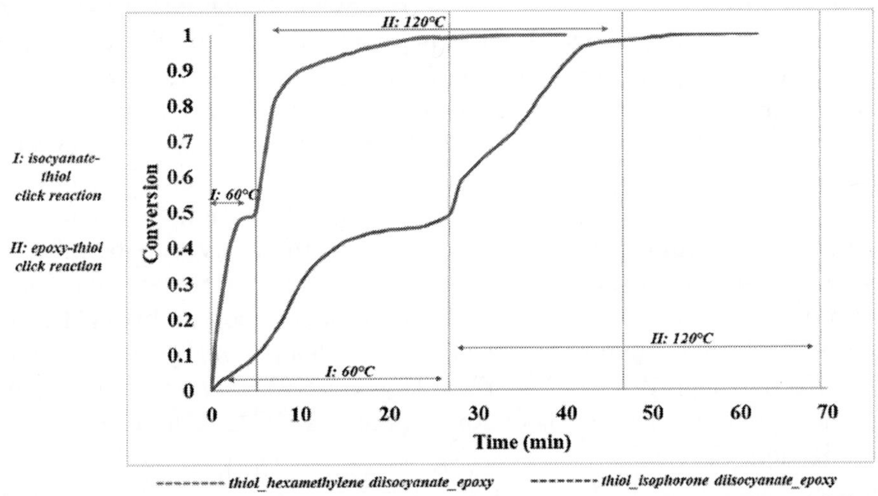

Fig. 1: Conversion plot of the dual curing systems (diisocyanate/diepoxy 0.5 molar ratio) determined by FTIR at 60°C (thiol-isocyanate) and at 120°C (thiol-epoxy)

Acknowledgements. The authors would like to thank MCIU (Ministerio de Ciencia, Innovación y Universidades) and FEDER (Fondo Europeo de Desarrollo Regional), MAT2017-82849-C2-1-R and -2-R) and to Generalitat de Catalunya (2017-SGR-77) for the financial support.

References

1. Konuray O.; Fernández-Francos, X.; Ramis, X.; Serra, A. Polymers **10**, 178 (2018)
2. Belmonte A.; Fernández-Francos, X.; De la Flor, S.; Serra A. Mech. Time-Depend. Mater.,21, 133 (2017).

RAFT POLYMERIZATION AS A TOOL FOR OBTAINING BIOMEDICAL MATRICES ON THE BASE OF N-VINYL-SUCCINIMIDE AND 5-VINYLTETRAZOLE

A. I. Gostev[1], D. M. Krygina[2], E. V. Sivtsov[1], V. A. Ostrovskii[1]

[1] *Saint-Petersburg State Institute of Technology, Saint-Petersburg, Russia*
[2] *Institute of Macromolecular Compounds of RAS, Saint-Petersburg, Russia,* krygina.dm@gmail.com

Vinyl tetrazoles are a special class of vinyl monomers. Polymers based on them are interesting as high-energy materials and have great potential for biomedical applications. The traditional radical homo- and copolymerization of these monomers has been studied for the last several decades, but there is no information about their polymerization using technics of controlled radical polymerization. In the present study, RAFT (reversible addition-fragmentation chain transfer) homopolymerization of 5-vinyltetrazole (5VT) was carried out and its copolymerization with N-vinylsuccinimide (VSI) in the presence of dibenzyltrithiocarbonate (BTC) as a RAFT agent. Copolymers of 5VT with VSI are promising as a base for obtaining new, water-soluble and non-toxic matrices for immobilizing low molecular weight substances, including physiologically active ones, as shown in the following Scheme:

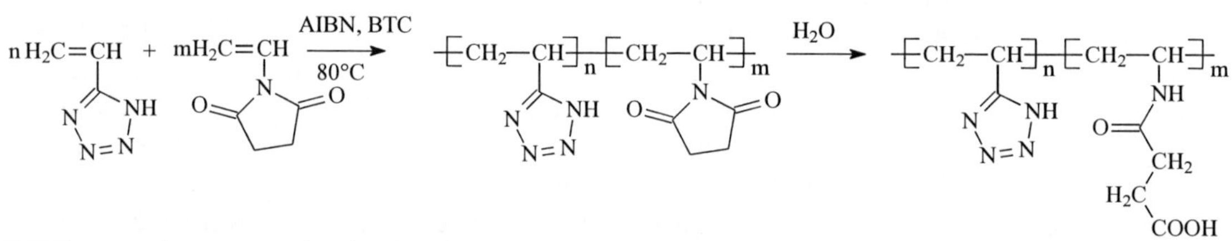

BTC was chosen on the basis of numerous data on its effectiveness as a RAFT agent with respect to vinyl monomers of various structures. The main purpose of controlled radical polymerization is to obtain polymers with a certain molecular mass (MM), narrow MM distribution and high compositional homogeneity in the case of the copolymerization of monomers, which differ greatly in their reactivity ratios [1]. Achievement of these goals is possible with efficiency of RAFT agent sufficient for the satisfactory control. As expected, due to the pronounced electron-withdrawing nature of the substituent in 5VT, the efficiency of BTC in its homopolymerization is low ($C_{tr} \sim 1.5$). However, at $[BTC]/[5VT] = 2 \cdot 10^{-2}$, $[AIBN]/[5VT] = 2 \cdot 10^{-3}$, the linear dependence of MM on the monomer conversion and the formation of polymers with narrow MM distribution were observed. When the [BTC]/[5VT] ratio decreases to $1 \cdot 10^{-2}$, the control is lost. Taking into account the above a known practical trick has been used: the oligomeric poly(5VT) was synthesized at a high BTC concentration and then was used as a new oligomeric RAFT agent with a much higher value of C_{tr}.

BTC is more effective in polymerization of VSI ($C_{tr} \sim 19$) [2], which is a more active monomer than 5VT in their copolymerization. Therefore, in this case BTC is more effective than in 5VT homopolymerization: $C_{tr} \sim 5.5$.

Acknowledgements. This work was supported by the Ministry of Science and Education of the Russian Federation (11.5362.2017/8.9).

References

1. E.V. Chernikova, E.V. Sivtsov. Polymer Science, Series B, **59**, 117-146 (2017).
2. E.V. Sivtsov et al. Polymer Science, Series B, **57**, 110-119 (2015).

NANOSCALE MAPPING OF INTERPHASE WITH AFM: POLYMER BASED NANOCOMPOSITES

I. Dobryden[1], T. Tokarski[1], M. Cortes Ruiz[2], G. Li[1] and Per M. Claesson[1,3]

[1] *Department of Surface and Corrosion Science, KTH Royal Institute of Technology, Stockholm, Sweden, illia@kth.se*
[2] *Department of Chemical Engineering, Grove School of Engineering, New York, United States*
[3] *RISE Research Institutes of Sweden, Division of Bioscience and Materials, Stockholm, Sweden*

Polymeric nanocomposites are an interesting material class within the developing field of nanotechnologies, and with a strong link to modern industry needs. Nanocomposite properties can be tuned in a controlled manner by introducing various filler nanoparticles possessing excellent material properties. Bulk nanomechanical and surface robustness are among the important properties to be improved by introducing filler nanoparticles. It was recently demonstrated, and is the state-of-the-art within this field, that dramatic increase in polymer nanocomposite mechanical properties mainly occurs due to the interfacial region and not due to nanoparticles themselves. This interfacial region represents a transitional volume between the nanoparticles and the polymer matrix, and it is often referred to as the interphase. Such interphase volume is becoming the dominating volume at the nanofiller diameter further decreases. Nanocomposites based on polymeric blends with different thermal properties can demonstrate even more complicated impact on the interphase volume, leading to more difficult evaluation. This was previously demonstrated in our work with a thermoplastic mixture of (poly(ethyl methacrylate) (PEMA) and poly(isobutyl methacrylate) (PiBMA)) with added silica nanoparticles studied with AFM[1]. The evaluation of interphase properties, from both elastic and viscoelastic nanocomposites, requires utilization of new developments within the multifrequency AFM family such as intermodulation AFM. Such measurements were previously conducted in our group and we evaluated that both elastic and viscoelastic surface responses of the interphase [2]. Thus, it was shown that atomic force microscopy methods are very valuable to probe such nanoscale material properties.

One of the remaining challenges in the field is the difficulty of determining the exact immersion depth of the nanoparticles found at the surface, which can give rise to false interpretation of AFM results, as we reported in [2]. To overcome these issues a special nanocomposite preparation is required. The polydimethylsiloxane (PDMS) polymer matrix and carboxylic modified polystyrene nanoparticles were chosen to prepare such nanocomposite. The carboxylic groups density on the particle surface defines a contact angle at the air-water interface, and a desired nanoparticle immersion depth can be achieved. Nanocomposites prepared according to this modified preparation procedure were obtained. The volumetric interphase phenomenon can be clearly identified and visualized for the first time. Moreover, nanocomposite surface toughness and correlation between interphase and nanoparticle removal during nanowear tests were investigated.

References

1. H. Huang et al., Composites Science and Technology, **150**, 111 (2017).
2. H. Huang et al., Journal of Colloid and Interface Science, **494**, 204 (2017)

SEC OF POLYMER WITH COMPLEX DISTRIBUTION OF MOLECULAR WEIGHT AND BRANCH-POINTS

M. Netopilík

Institute of Macromolecular Chemistry, Heyrovský Sq. 2, 162 06 Prague 6, Czech Republic,
netopilik@imc.cas.cz

It is generally believed that polymer molecules are separated in size exclusion chromatography (SEC) according to their hydrodynamic volume. However, the elution behaviour of branched[1] and brush[2] polymers deviates in some cases from that of linear-chain polymers. The usual decrease in molecular weight, M, and root-mean-square radius, $\langle s^2\rangle^{1/2}$, with increasing elution volume is then followed by an abrupt increase in both mean values of $\langle s^2\rangle^{1/2}$ and, M, in the tail of the elution curve, where a decrease is expected in the analogy with the linear-chain molecules.

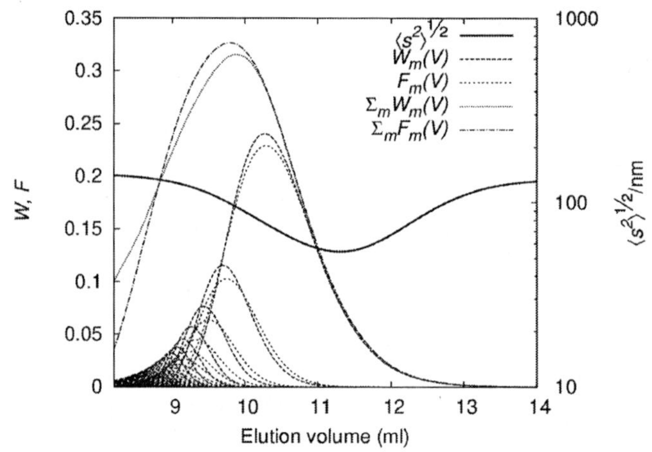

Fig. 1. Comparison of non-broadened and broadened elution curves of fractions with m branch points, $W_m(V)$ and $F_m(V)$, respectively, and their sums, with calculated values of root-mean-square radius $\langle s^2\rangle^{1/2}$.

We modeled the analysis using the Stockmayer theory in modification of 'Thurmond and Zimm[3] and the assumption of the capture of the molecules followed by the release of the entangled molecule in dependence on the number of the free chain ends, i.e., on the number of the tetrafunctional branch points. An example of the results of the model SEC analysis is in Fig. 1.

Acknowledgments: The financial support of the Grant Agency of the Czech Republic (project 17-04258J) is gratefully acknowledged.

References

1. Š. Podzimek, T. Vlček, C. Johann, J. App. Polym. Sci., **81**, 1588-1594 (2001).
2. M.Gerle, K.Fischer, S.Roos, A.H.E. Müller, M.Schmidt, S.S. Sheiko, S.Prokhorova, M.Möller, Macromolecules, **32**,2629-2637 (1999).
3. M. Netopilík, P. Kratochvíl, Polym. Int. **55**, 196-203, (2006)

GRAPHENE OXIDE BASED POLY(VINYL ALCOHOL) NANOCOMPOSITE FILMS: CONTROL OF MECHANICAL PROPERTIES

T. Panova, A. Efimova, A. Berkovich, A. Efimov

Polymer Department, Faculty of Chemistry, M.V. Lomonosov Moscow State University, Moscow, Russia, tvk@genebee.msu.ru

Poly(vynil alcohol) (PVA) is a semi-crystalline polymer readily soluble in water with excellent film-forming properties. Numerous fillers have been used for the preparation of PVA nanocomposites, among which graphene oxide (GO) have attracted a great attention. GO, similar to graphene, has remarkable mechanical, electrical and thermal properties. GO also has a considerable amount of hydrophilic oxygen-containing groups, such as epoxide and hydroxyl groups, across the sheet, and carbonyl and carboxyl groups at the edges. These hydrophilic groups can change van der Waals interactions in graphene layers and provide high value of dispersibility of GO in water, which allows to manufacture composites using aqueous solutions, without using toxic solvents. The interactions of hydrophilic groups of GO with hydroxyl groups in PVA provide an excellent compatibility of the filler with the polymer matrix in the resulting composite.

In this work composite films of PVA filled with different amounts of GO were prepared by solution casting technique, and their morphology, physical and mechanical properties were studied. GO was found to be uniformly distributed in the resulting composite, which proves the excellent compatibility of the filler with the polymer matrix. GO incorporation does not affect significantly the degree of crystallinity of PVA, but increases the thermal stability of the polymer. For PVA/GO composites, the increase of Young`s modulus and yield stress were found with an increase of GO content. However, the addition of GO led to the sharp decrease of elongation at break. In order to improve the plasticity of composite films, their modification with cold rolling was performed. It was shown that films cold rolling results in transition from brittle to ductile state of the samples. Another way of the modification of films plasticity used was the addition of glycerol as a plasticizer to the PVA/GO composite films. PVA based composite films with different amount of GO and glycerol were studied. The addition of certain amount of glycerol to the PVA/GO composite leads to the simultaneous Young`s modulus and yield stress decrease, and elongation at break increase, i.e. the composite films became much more ductile. Since water also can serve as a plasticizer for PVA, the mechanical properties of PVA/GO composites with glycerol containing different amounts of water will be also discussed.

Acknowledgements. This work was supported by Russian Science Foundation (project no. 17-73-20266).

INFLUENCE OF ALIPHATIC MOIETIES IN DIISOCYANATES ON CHAIN EXTENSION KINETICS OF ADIPATE MACRODIOLS

S. Mačiulytė, A. Strakšys, S. Asadauskas

Department of Chemical Engineering and Technologies, Center for Physical Sciences and Technology, Vilnius, Lithuania, asadauskas@chi.lt

Plastics, adhesives, resins and other polyurethane (PUR) formulations are widespread in various industrial, food, biomed and other applications [1]. In many cases two component blends are used, where "OH-component" represents a viscous prepolymer, which can be crosslinked with isocyanates into fully cured PUR. Adipate macrodiols are often employed in manufacture of non-yellowing and other high performance PUR. Their chain extension must be accurately controlled in order to obtain a needed molar mass distribution [2]. In this study, an adduct of adipic acid with ethylene glycol was used as an ester type macrodiol, while an adduct of adipic a. with diethylene glycol as an ether-ester type macrodiol, Fig. 1.

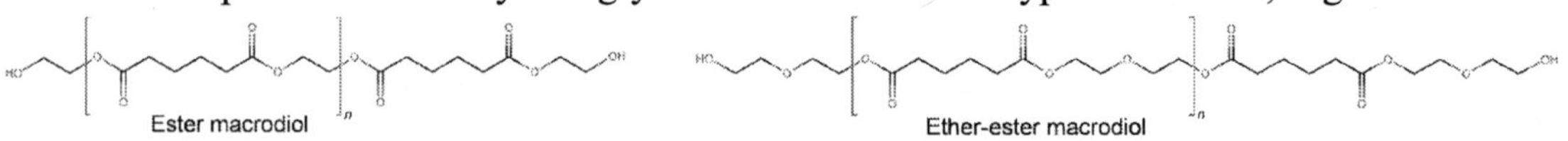

Fig. 1. Structures of adipate microdiols (n ~ 12-14), used for chain extension with diisocyanates

Chain extension of excess macrodiols was performed at 2:1 and 3.33:1 molar ratios with respect to aliphatic diisocyanates: either hexamethylene diisocyanate (HDI) or isophorone diisocyanate (IPDI). Titration of polymerization mixture aliquots with dibutyl amine was exercised to monitor the amount of unreacted NCO (isocyanate) groups, Fig. 2.

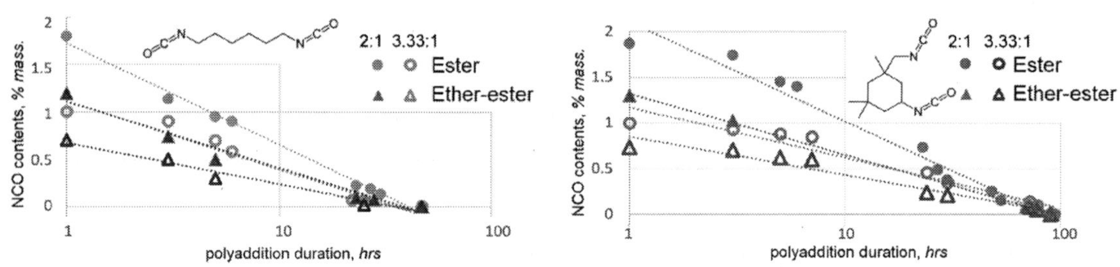

Fig. 2. Isocyanate (NCO- group) depletion at 80°C during chain extension of adipate macrodiols in 2:1 or 3.33:1 molar excess with aliphatic diisocyanates HDI (left) and IPDI (right).

Isocyanate depletion appears to follow approximate semilog dependence on the reaction duration, as might be inferred from satisfactory least square fits. R^2 values exceed 0.95 for all chain extensions, except for IPDI at 1:3.33 ratio to the ester macrodiol. Overall, IPDI seems to react slower than HDI, most likely due to steric hindrance, although the presence of one secondary alpha α-C atom might also affect the reactivity of NCO group in IPDI. Established semiquantitative trends might be beneficial for better control of prepolymer production in industrial manufacturing.

Acknowledgment. This study was carried out under project TERMINUS, funded by the European Union under Horizon 2020. Call: H2020-NMBP-ST-IND-2018. Grant Agreement: 814400.

References

1. B. N. Rao, P. J. P. Yadav, K. Malkappa, T. Jana., Polymer, **77** (2015) 323-333.
2. N. Akram, K. M. Zia, M. Saeed, M. Usman, S. Saleem, J. Appl. Polym. Sci., **135** (2018) 46635.

CHARACTERIZATION OF THE FUNCTIONAL PROPERTIES OF A NEW CLASS OF VITRIMERS BASED ON POLY(THIOURETHANE) NETWORKS

F. Gamardella[1], F. Guerrero[1], S. De la Flor[2], X. Ramis[3], À. Serra[1]

[1] *Dept. of Analytical and Organic Chemistry, Universitat Rovira i Virgili, C/ Marcel·lí Domingo, Edif. N4. 43007, Tarragona, Spain*
[2] *Dept. of Mechanical Engineering, Universitat Rovira i Virgili, Av. Països Catalans, 26, 43007 Tarragona, Spain, Silvia.delaflor@urv.cat*
[3] *Thermodynamics Laboratory, ETSEIB Universitat Politècnica de Catalunya, Av. Diagonal, 08028, Barcelona, Spain*

Vitrimers are a new promising class of polymeric materials which are attracting increasing attention thanks to their thermosetting characteristics, but with capability of being reprocessed and recycled, which make them more environmentally friendly in reference to the conventional thermosets. These materials exhibit an interchange of covalent bonds above a certain temperature called the freezing topological temperature (T_v) that enables self-healing and recycling as well as the relaxation of residual stresses [1, 2]. Shape memory materials based on vitrimers overcome limitations of the traditional thermosets enabling the production of complex temporary shapes and the formation of sophisticated and geometrically complex forms, thanks to the combination of elasticity (shape memory) and plasticity (vitrimers). In addition, these materials, after recycling, can also exhibit comparable mechanical performance [3].

In this study, we present a new class of vitrimeric materials consisting in poly(thiourethane) networks that can be easily produced from commercially available precursors. The reshaping, self-welding and stress relaxation capabilities are based on the trans-thiocarbamoylation reaction. A complete thermal and mechanical characterization is carried out at temperatures above and below T_v in order to totally control the shape memory performance, the self-healing capability and self-welding ability.

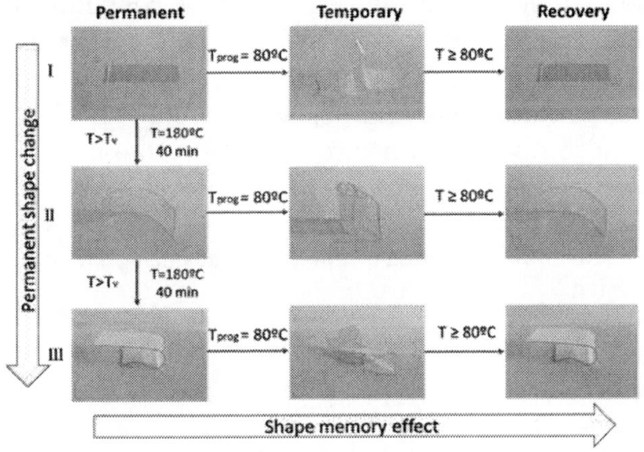

Fig. 1. Qualitative demonstration of shape memory behavior and permanent/plastic shape change for the vitrimeric material.

Acknowledgements. The authors would like to thank MCIU (Ministerio de Ciencia, Innovación y Universidades) and FEDER (Fondo Europeo de Desarrollo Regional) (MAT2017-82849-C2-1-R and MAT2017-82849-C2-2-R) and Generalitat de Catalunya (2017-SGR-77) for the financial support.

References

1. D. Montarnal et al., Science, **334**, 965–968 (2011).
2. W. Denissen et al., Chem. Sci., **7**, **30**–38 (2016).
3. Q. Zhao et al., Sci Adv., 2, e1501297 (2016).

IMPACT OF FILM MORPHOLOGY ON ELECTRONIC STRUCTURE IN THE PROTOTYPICAL SEMICONDUCTING POLYMER POLY-3-HEXYLTHIOPHENE

A. Kadashchuk[1,2], A. Vakhnin[1], A. Zhugayevych[3], A. Köhler[4]

[1] *Institute of Physics, National Academy of Sciences of Ukraine, Kyiv, Ukraine, kadash@iop.kiev.ua*
[2] *IMEC, Kapeldreef 75, Leuven, Belgium*
[3] *Center for Electrochemical Energy Storage, Skolthech, Moscow, Russia*
[4] *Bayreuth Institute of Macromolecular Research (BIMF), University of Bayreuth, Bayreuth, Germany*

Semiconducting π-conjugated polymers are currently employed in organic solar cells and organic field-effect transistors mainly as they allow for a low-cost and low-temperature device fabrication compatible with flexible plastic substrates. It is commonly believed that charge transport in conjugated polymers is controlled by thermally activated hopping between segments of the same polymer chain separated from each other by topological defects and/or by carrier jumps between adjacent segments of different chains. The length of segments is subject to random variations that, together with random potential fluctuations caused by structural disorder, is the major source of the energy disorder in these polymers, implying a relatively broad density-of-states (DOS) energy distribution for neutral and charged excitations. The intrinsic energy disorder along with possible extrinsic traps are major factors limiting charge transport in these materials. Thus, an accurate picture of the density of states associated to traps and energetic disorder in such films is important as these have a strong impact on that aspect of the device performance which is related to charge carrier mobility.
Poly-3-hexylthiophene (P3HT) polymer is an archetypal organic semiconductor commonly used in different optoelectronic devices. Despite the P3HT is one of the most studied and technologically developed polymeric semiconductor, its structure-property relationships is still not fully understood. P3HT can adopt different morphologies and depending on solvent, temperature, molecular weight, and sample preparation, it can exist in an amorphous phase with coiled chain conformations or in an aggregated phase containing planarized chains prone to form semicrystalline domains. In this work we present a comprehensive spectroscopy and charge trapping studies using the low-temperature thermally stimulated luminescence technique in P3HT films of various polymorphs and morphologies. As a result, by combining several approaches we have resolved structure of semicrystalline P3HT and established new relationships between morphology and electronic properties.

Acknowledgements. This research was supported by bilateral Lithuanian-Ukrainian cooperation project № M/9-2019 and by the VW Foundation.

DEVELOPMENT OF A MOLECULARLY IMPRINTED POLYMER-BASED SENSOR FOR ELECTROCHEMICAL DETECTION OF MACROLIDE ANTIBIOTICS

A. G. Ayankojo, J. Reut, A. Öpik, V. Syritski

Department of Materials and Environmental Technology, Tallinn University of Technology, Tallinn, Estonia, akinrinade.ayankojo@taltech.ee

Antibiotic water pollutants are gaining increasing popularity among the list of environmental pollutants owing to their wide usage in both human and veterinary medicine. Macrolides are widely used antibiotics possessing potency against broad bacteria strains. Determination of macrolides especially erythromycin (Ery), clarithromycin (Clari) and azithromycin (Azi) in water body of many countries indicates their entry into the environment through various route hence, their inclusion in the EU watch list of environmental pollutants. Persistence of antibiotics in environmental water could affect biological activities but has a more detrimental effects of enhancing the development of virulent strains of pathogenic bacteria[1], thus to ensure the safety of life while preserving environmental sanctity, monitoring levels of these molecules in aqueous environment is of utmost importance.

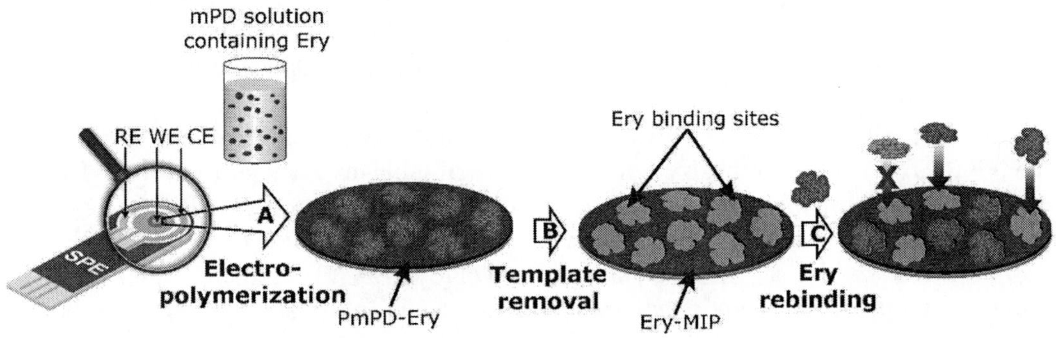

Fig. 1. Schematic representation of the protocol for Ery-MIP/SPE synthesis..

Although, traditional analytical techniques such as high performance liquid chromatography (HPLC) and solid phase extraction have demonstrated remarkable success, they suffer drawbacks in their huge size, cost and sophisticated operational principles. Therefore, developing portable and selective sensing devices for detecting these antibiotics constitutes a developing but urgent concern. In this regard, combining a compact but low cost screen printed electrochemical (SPE) sensor with a selective recognition layer such as a molecularly imprinted polymer (MIP) could suggest a potential solution[2]. Hence, a major area of our research entails fabricating a macrolide antibiotic selective MIP film integrated with the working electrode of SPE, as a portable sensor for identifying these pollutants in water. This report demonstrates the possibility of preparing a portable MIP based SPE sensor by employing Ery as a representative macrolide antibiotic (Fig.1). Sensor optimization and performance characterization indicates its suitability for use in target media thereby presenting the sensor as a potential monitoring platform for the desired selective and cost effective detection of macrolides in water.

Acknowledgements. This work was supported by Estonian Research Council (grant PRG307).

References

1. G. Ayankojo et al., Biosensors and Bioelectronics, **118**, 102–107 (2018).
2. Hayat & J. Marty, sensors, **14**, 10432-10453 (2014).

MOLECULARLY IMPRINTED POLYMERS AS SYNTHETIC ANTIBODIES FOR NEUROTROPHIC FACTOR PROTEINS DETECTION

A. Kidakova, R. Boroznjak, J. Reut, A. Öpik, V. Syritski

Department of Materials and Environmental Technology, Tallinn University of Technology, Ehitajate tee 5, 19086 Tallinn, Estonia, anna.kidakova@ttu.ee

Today label-free detection of proteins has become of extensive demand in fundamental research as well as in clinical practice providing an alternative to the widely used label-based ELISA methods. Most of the current biosensing systems utilize labile biological recognition elements, such as DNA, antibodies, or enzymes that can capture a target analyte with very high affinity and specificity. However, due to limited operating conditions and the high cost of production of biological receptors, there has been significant interest in artificial systems capable of mimicking natural receptors, such as molecularly imprinted polymers (MIP). MIPs have been shown to possess several advantages as an alternative recognition material for chemosensor[1].

Due to the increase in the incidence of neurological and mental disorders, the research devoted to the early detection of biomarkers of such diseases has become popular. Neurotrophic factor (NF) proteins are a family of proteins secreted from neurons and neuron-supporting cells and were found to be associated with several neurological diseases such as Alzheimer, Parkinson's and mental disorders[2].

In this study, MIP films for selective detection of brain-derived NF (BDNF) and cerebral dopamine NF (CDNF) were synthesized on the surface of label-free sensor platforms. BDNF-MIP was synthesized on the gold surface of screen-printed electrodes (SPE) by surface-initiated controlled/living radical photopolymerization of acrylamides[3]. CDNF-MIP was reliably interfaced with a surface acoustic wave (SAW) sensing platform by electropolymerization of m-phenylenediamine on the SAW chip surface with previously immobilized target protein. The binding properties, detectability and selectivity of the BDNF-MIP SPE sensing devices toward BDNF were determined using differential pulse voltammetry (DPV). The synthesized CDNF-MIP films were characterized by monitoring the binding kinetics with the SAW system. Both BDNF-MIP and CDNF-MIP sensors were capable of detecting the target protein and could recognize it in mixtures containing proteins of different molecular weight. The proposed synthesis methods of NF- selective MIP film integrated to a label-free sensing platforms can be a premise for fabrication of robust, cost-effective sensors delivering faster and more error-free detection of the respective compounds than current enzyme-based systems and provides new diagnostic tools for neurodegenerative diseases, where presymptomatic diagnostics does not exist.

Acknowledgments. This work was supported by Estonian Research Council (grant PRG307).

References

1. P.S. Sharma et al., Analytical and Bioanalytical Chemistry, 402(10), 3177–204.
2. M. Lindahl et al., Neurobiology of Disease, 97(2017) 90-102.
3. A. Kidakova et al., P Est Acad Sci, 68(2019) 158-67.

TRAPPED STATES AFFECT TEMPERATURE-DEPENDENT INTERFACIAL PROPERTIES OF PIPOZ-CONTAINING POLYMERS

J. An[1], A. Dėdinaitė[1,2], F. M. Winnik [3,4,5] and P. M. Claesson[1,2]

[1] *KTH Royal Institute of Technology, School of Engineering Sciences in Chemistry, Biotechnology and Health, Department of Chemistry, Surface and Corrosion Science, Drottning Kristinas väg 51, SE-10044 Stockholm, Sweden*
[2] *RISE Research Institutes of Sweden, Materials and Surfaces, SE-114 86 Stockholm, Sweden*
[3] *Department of Chemistry and Faculty of Pharmacy, University of Montreal, CP 6128 Succursale Centre Ville, Montreal, QC H3C3J7, Canada*
[4] *WPI International Center for Materials Nanoarchitectonics (MANA), National Institute for Materials Science, 1-1 Namiki, Tsukuba, Ibaraki 305-0044 Japan*
[5] *Department of Chemistry and Faculty of Pharmacy, University of Helsinki, Helsinki Finland*

Poly(2-isopropyl-2-oxazoline) (PIPOZ) is a thermoresponsive polymer that in aqueous solutions phase separates at above ≈46 °C. Diblock and pentablock co-polymers of PIPOZ and the cationic poly(3-acrylamidopropyl-trimethylammonium) (PAMPTMA) have been prepared and their temperature-dependent properties on silica surfaces were evaluated. The cationic block provides anchoring of a polymer to the negatively charged silica surfaces. The polymer adsorption increases with increasing temperature due to worsening of the solvent quality. Desorption upon cooling is limited, giving rise to significant adsorption hysteresis. Thus, long-lived trapped non-equilibrium states are prevalent.
Surface and friction forces between ***pre-adsorbed layers*** of such copolymers were explored at different temperatures. We found that with worsening of the solvent condition an attraction developed between the adsorbed PIPOZ layers, and the friction forces increased. With polymers present in solution the trapped states were common and the adsorbed amount was dependent on the temperature history. In this case the surface forces were purely repulsive at temperatures < 40 °C. A local force minimum at short separation developed at 40 °C and a strong attraction due to capillary condensation of a polymer-rich phase was observed close to the bulk phase separation temperature. Further, the friction forces were found to decrease with increasing temperature due to the increase in adsorbed amount, and also friction forces were dependent on temperature history. Clearly, trapped states are of importance whenever kinetics is slow, and these can have profound effects on properties in complex formulations.

References
1. An, J., Xiaoyan, L., G., Dėdinaitė, A., Korchagina, E. , Winnik, F.M., Claesson, P.M. "Effect of solvent quality and chain density on normal and frictional forces between electrostatically anchored thermoresponsive diblock copolymer layers", JCIS, 487, 88–89, (2017)
2. An, J., Xiaoyan, L., G., Linse, P., Dėdinaitė, A., Winnik, F.M., Claesson, P.M. "Tethered poly(2-isoporopyl-2-oxazoline) chains: Temperature effects on layer structure and interactions probed by AFM experiments and modelling", Langmuir, 31, 3039-3048 (2015).

INFLUENCE OF THE ADHESION ON THE MECHANICAL CHARACTERISTICS OF SELF-REINFORCED POLYMER-VENEER-COMPOSITES

C. v. Boyneburgk, P. Sälzer, J. Fuchs, H.-P. Heim

Institute of Material Engineering, Polymer Engineering, University of Kassel, Kassel, Germany, c.vonboyneburgk@uni-kassel.de

Three-layered composites were manufactured based on self-reinforced polypropylene and wood veneer (European beech, *Fagus sylvatica*) in a high precision hot compaction process. The outer layers each consist of wood veneer with a thickness of 1 mm, the middle layer of eight layers self-reinforced thermoplastic fabric, each with a thickness of 0.18 mm. In order to produce additional melt for an improved physical adhesion, thin polymer films were produced using a flat film line, one charge exclusively with polypropylene, the other charge made of polypropylene with 5 wt. % maleic anhydride grafted polypropylene wax as coupling agent. First investigations showed that the use of maleic anhydride as a coupling agent leads to the expected enhanced mechanical characteristics under quasi-static as well as dynamic load. In order to determine the deeper connections between the use of the coupling agent and the adhesion as well as the resulting mechanical properties in the composite, further investigations were carried out. First, differences in the penetration depth of the thermoplastic material into the wood could be clearly determined by analysis of light microscopy and μ-CT images. It could be observed that polypropylene with coupling agent penetrates far less deeply into the wood structure than polypropylene without coupling agent.

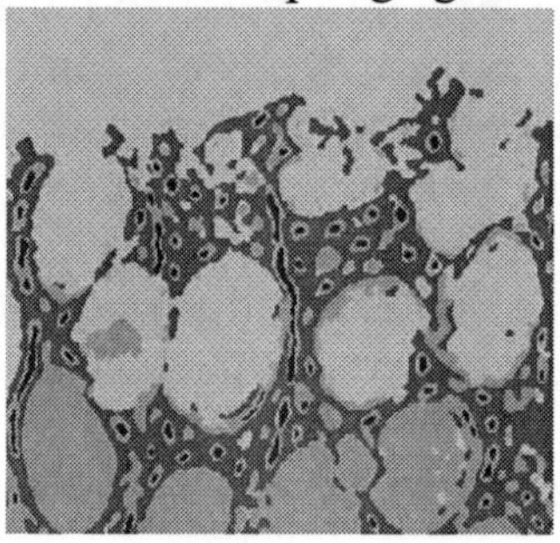

Fig. 1. μ-CT image of the interface between wood and polymer

The assumption that this was caused by an increased viscosity because of the maleic anhydride could be disproved by high-pressure capillary rheometer measurements, which showed a lower viscosity for the polypropylene containing maleic anhydride. Further investigations using FTIR analysis and contact angle measurements showed that the surface energy of the used films was enhanced throughout the use of maleic anhydride. The present results suggest that the use of maleic anhydride renders the adhesion of the polymer to the cell wall whereby the ports of entry into the wood structure close in short time and make further polymer penetration impossible. Despite this fact, the use of maleic anhydride leads to improved mechanical properties. This allows the conclusion that the effect of the chemical adhesion of maleic anhydride on the cell walls overlies the physical adhesion by undercuts so strongly, that even the higher penetration of pure polypropylene into the wood structure cannot produce improved properties.

NEW SILVER NANOCOMPOSITES WITH CYTOTOXIC ACTIVITY

M. N. Gorbunova, D. V. Eroshenko

Institute of Technical Chemistry, Ural Branch of Russian Academy of Sciences, Korolev str., 3, Perm 614013, Russia, mngorb@yandex.ru

The rapid development of the chemistry associated with the nanoscale state of matter led to a significant increase in the number of investigation devoted to the synthesis and properties of silver nanoparticles [1]. The incorporation of nanoparticles into various matrices extend their utility in material and biological applications. It is known from the literature that the stability of the particles is significantly affected by the nature of the stabilizing matrix. Our work is focused on incorporating of silver nanoparticles into matrice of new copolymers based on N,N-diallyl-N'-acylhydrazines.

The copolymers of N,N-diallyl-N'-acetylhydrazine (DAAH), N,N-diallyl-N'-propanoylhydrazine (DAPH), N,N-diallyl-N'-butanoylhydrazine (DABH) and N,N-diallyl-N'-benzoylhydrazine (DABEH) with N-vinylpyrrolidone have been obtained by free radical polymerization. N,N-diallyl-N'-acylhydrazines copolymerize with N-vinylpyrrolidone with formation of cis-, trans-stereoisomeric pyrrolidine structures in cyclolinear polymer chain.

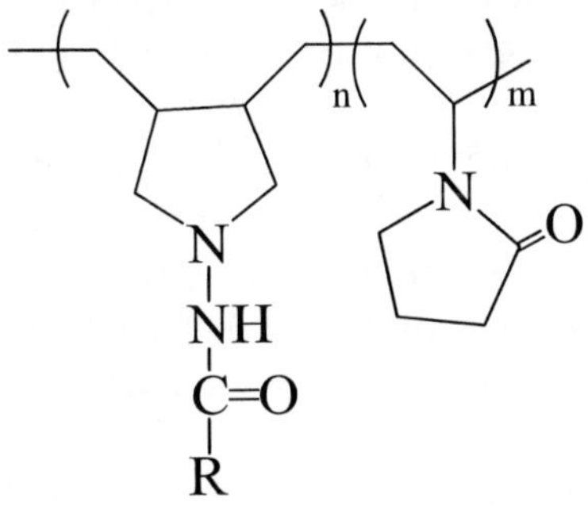

$R = CH_3, C_2H_5, C_3H_7, C_6H_5$

Synthesis of silver nanocomposites was conducted by the reduction of $AgNO_3$ with $NaBH_4$ in solution of copolymer. Reaction proceeds via formation of the stable dark brown sols, from which nanocomposites were separated.

SEM results prove the obtaining of nanocomposites with regular narrow-dispersed distribution of silver nanoparticles in polymer matrice. Nanoparticles of spheric and elliptic forms were obtained. The average particle sizes of silver nanoparticles are in the range of 13 to 22 nm. Investigation showed that new nanocomposites have a significant cytotoxic activity and can be used for medical applications.

Acknowledgement: Financial support by the Russian Foundation for Basic Reseach (grant № 19-43-590019-r_a) is gratefully acknowledged.

References:

1. D.D. Evanoff, G. Chumanov, ChemPhysChem, **6**, 1221- 1231(2006)

NONDESTRUCTIVE ADSORPTION OF LIPOSOMES ON THE SURFACE OF LATEXES

O. V. Zaborova[1], M. N. Chernikova[2], P. N. Veremeeva[1]

[1] *Department of Chemistry, Lomonosov Moscow State University, Moscow, Russia,* olga.zaborova@gmail.com
[2] *D.I. Mendeleev University of Chemical Technology of Russia, Moscow, Russia*

Liposomes are the spherical vesicles formed from natural and synthetic lipids ranging in size from 20 nm to several micrometers. Due to their size and amphiphilic character, liposomes represent the perspective systems for delivery of the biologically active substances. There are several liposomal formulations on the market based on the use of individual, so-called "conventional", liposomes. The use of conventional liposomes has several disadvantages such as small internal volume of the container, low mechanical and aggregative stability, etc.
Modification of liposomes with polyelectrolytes allows to overcome some above-mentioned disadvantages, but the formation of the polymer/liposome complex leads to undesirable aggregation in the system. In addition, adsorption of polyelectrolytes on vesicles causes a number of structural reorganization in the lipid membrane that can lead to a disruption of liposomal membrane. Adsorption of liposomes on a surface allows to control the quantity of the adsorbed liposomes, but the interaction of liposomes with a surface leads to the partial destruction of liposomes.
One of the most successful approaches for concentrating of liposomes on a surface is electrostatic adsorption of anionic liposomes on the cationic colloidal particles (polystyrene core with grafted linear polycationic chains) [1]. This method allows to concentrate dozens of liposomes without the disruption of the liposomal structure. However, this approach assumes existence of a "soft" layer between a liposome and a surface, which introduces its limitations on the nature of the carrier surface.
Modification of a surface of solid particles is a rather difficult task, thus the purpose of this work is to investigate the possibility to use a "soft" layer on a liposome to prevent the contact with the surface Anionic latex particles with a size of 350 nm were used as a model surface. The electrostatic adsorption of liposomes on the latex was ensured by using the cationic (mole fraction of cationic lipid, $\nu_+ = 0.1$) or zwitterionic liposomes ($\nu_+ = \nu_- = 0.1$). The "soft" layer was formed by polyethylene glycol (PEG), thus the lipid modified with PEG was incorporated into liposomal membrane. The mole fraction (ν) of the PEGylated lipid in liposomes was varied from 0 to 0.2.
We showed that the introduction of PEGylated lipid into the liposomal membrane do not change the basic physico-chemical properties of liposomal adsorption on the latex surface. We found that with an insufficient amount of PEGylated lipid ($\nu = 0.05$, 0.05 and 0.025 for PEG 750, 2000 and 5000, respectively) and, also, an excess amount of PEGylated lipid ($\nu = 0.2$, 0.15, and 0.1 for PEG 750, 2000, and 5000, respectively), liposome destruction during adsorption was observed. Thus, we have shown that the embedding of intermediate amounts of PEGylated lipid into the membrane forms the "soft" layer and helps to prevent the destruction of liposomes while adsorbed on a latex surface.

Acknowledgements. The reported study was funded by RFBR according to the research project № 18-33-00591.

References

1. A.A. Yaroslavov et al, Adv. Colloid Interface Sci., **226**, Part A, 54–64 (2015).

NANOCONTAINERS BASED ON CHITOSAN AND ANIONIC LIPOSOMES

A. Efimova[1], G. Krivtsov[2], N. Melik-Nubarov[1], I. Grozdova[1], A. Yaroslavov[1]

[1] *Polymer Department, Faculty of Chemistry, M.V.Lomonosov Moscow State University, Moscow, Russia, ephimova@genebee.msu.su*
[2] *Mechnikov Research Institute of Vaccines and Sera, Moscow, Russia*

For several decades liposomes have been used as containers for encapsulation and delivery of drugs. Recently, multi-liposomal containers have been described, composed of dozens of liposomes with different fillings at variable ratios. This allows the enhanced drug loading and offers a simple technology for fabrication of patient-oriented multi-functional drug formulations.
In this work nanocontainers were prepared via a direct electrostatic adsorption of two- and three-component anionic liposomes on the surface of chitosan nanoparticles. Chitosan is a commercially available cationic biopolymer of natural origin that is known for being biocompatible and biodegradable. This polymer is metabolised by certain human enzymes.
It was found that each chitosan particle could adsorb up to 150 anionic liposomes. The conjugates demonstrated stability towards dissociation in physiological media. At the same time, the conjugates degraded, being attacked by proteolytic enzymes, down to 10-15 nm particles, that was accompanied by a complete release of the encapsulated antitumor antibiotic doxorubicin into surrounding solution. It was found that chitosan-liposome complex and the products of its biodegradation were non-toxic in a wide range of concentrations. The size of the conjugates (250-400 nm) allows to recommend them as a potential carrier of biologically active substances capable of penetrating into the cells by means of the passive transport mechanism.
Thus, the findings of this work seem to be promising in the field of drug delivery.

Acknowledgements. This work was supported by Russian Foundation for Basic Research (project 19-03-00314.

POLYPROPYLENE REINFORCED WITH SHORT PET FIBRES – MANUFACTURING ISSUES AND IMPACT BEHAVIOUR

P. Franciszczak[1], A. Błędzki[1], E. Pleskot[1], M. Radwański[2], A. Kovalovs[3], G. Japins[3], K. Kalnins[3], A. Chate[3], R. Merijs-Meri[4], J. Zicans[4]

[1] *Institute of Materials Science, West Pomeranian University of Technology Szczecin, Szczecin, Poland*
piotr.franciszczak@zut.edu.pl
[2] *Ekotex, Namysłów, Poland*
[3] *Institute of Materials and Structures, Riga Technical University, Riga, Latvia*
[4] *Institute of Polymer Materials, Riga Technical University, Riga, Latvia*

The development of manufacturing methodology for polypropylene reinforced with short PET fibres provided a composite with superior impact properties that can be processed by injection moulding. This study shows the crucial issues of PET fibres selection, fibre pellet fabrication and composite compounding, which can be easily up-scaled and implemented in the plastic industry. The impact performance of PP/PET fibres composites was evaluated using drop-weight tests and compared to unreinforced PP and composite counterparts reinforced with short glass and viscose fibres. The basic mechanical properties were also related to reference composites.

Acknowledgements. The study was done within the framework of the project 'High performance short-fibre biobased hybrid composites for injection moulding (HyBiCo)' supported by the European Social Funded m-era.net network project No 4297.

POLYPROPYLENE AND ORGANOCLAY NANOCOMPOSITE MECHANICAL PROPERITIES

M. Varkale[1], I. Bochkov[1], R. Merijs Meri[1], J. Zicans[1], T. Ivanova[1], A. K. Bledzki[2]

[1]*Institute of Polymer Materials, Riga Technical University, Riga, Latvia*
[2]*Institute of Materials Science, West Pomeranian University of Technology Szczecin, Poland*

Nanocomposites based on layered inorganic compounds such as clays are in widely used owing to their high mechanical strength and good thermal resistance.
Nanoparticles have an extremely high surface-to-volume ratio, which dramatically changes their properties when compared with their bulk-sized particles. Nanocomposites show improvement in mechanical properties (flexular strength, tensile strength, stiffness, and toughness). However, clay particles are polar, but most of the thermoplastic polymers, including polypropylene, are non-polar. Using a maleic anhydride compatibilizer can increase compatibility between the polar fibre and the nonpolar matrix, leading to improved functional properties of the composite material. The investigated maleic anhydride compatibilizer modified polypropylene composites possess increased tensile strength and Young's modulus.

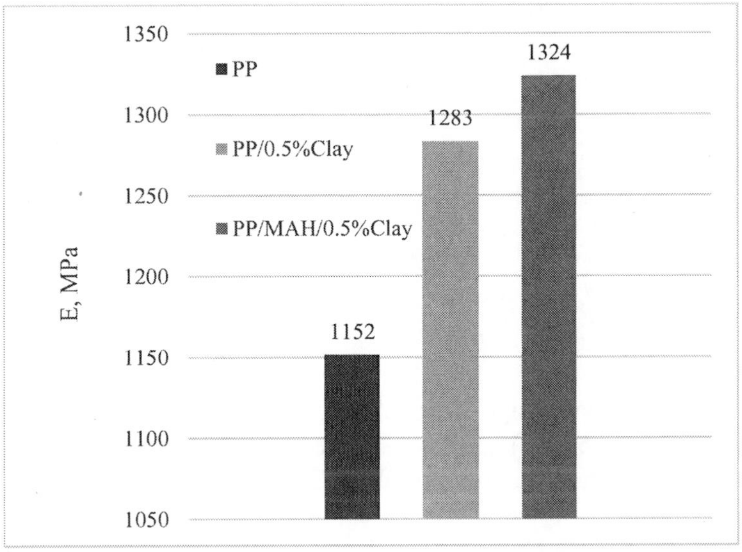

Fig. 1. Flexular modulus improvement by adding organoclay nanoparticles and maleic anhydride compatibilizer

Acknowledgements. This work has been supported by European Council HORIZON 2020 Program ERA.NET Cofund project "High performance short fibre biobased hybrid composites for injection moulding" (HyBiCo).

References

1. W. Lertwimolnun et al., "Influence of compatibilizer and processing conditions on the dispersion of nanoclay in a polypropylene matrix", Polymer, 46 (**10**), 3462–3471 (2005).
2. Oana M. Istrate **et al.,** **"**Structure and properties of clay/recycled plastic composites**"**, Applied Clay Science (**156**), 144-151 (2018).

POLYETHERS CONTAINING 4-(CARBAZOL-2-YL)-7-ARYLBENZO[C]-1,2,5-THIADIAZOLE CHROMOPHORES AS SOLUTION PROCESSED MATERIALS FOR HOLE TRANSPORTING LAYERS OF OLEDS

G. Kručaitė[1], D. Tavgenienė[1], Z. Xie[2], X. Lin[2], B. Zhang[2], S. Grigalevičius[1]

[1] *Department of Polymer Chemistry and Technology, Kaunas University of Technology, Kaunas, Lithuania, gintare.krucaite@ktu.lt*
[2] *State Key Laboratory of Polymer Physics and Chemistry, Changchun Institute of Applied Chemistry, Chinese Academy of Sciences, Changchun, China*

Solution processed derivatives containing substituted carbazole fragments are among the most studied materials for organic electronics due to their good chemical and environmental stability, and high positive charge mobility in their layers [1, 2,]. In addition, derivatives containing electronically isolated carbazole rings have high triplet energies and are widely used as host materials for electro-phosphorescent devices (PhOLEDs) [3].
The polymers (**P1**-**P2**) containing electronically isolated 4-(carbazol-2-yl)-7-arylbenzo[c]-1,2,5-thiazole fragments was carried out by a multi-step synthetic route and are shown in Scheme 1.

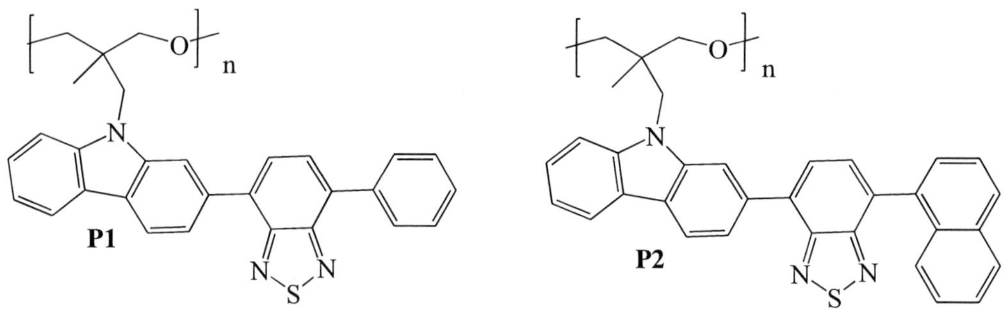

Scheme 1.

The amorphous polymers have high thermal stability as well as very high glass transition temperatures (154 °C and 163 °C). It was observed that layers of polymers have rather high and similar values of ionization potential of about 5.9 eV. The investigations demonstrate that the thin electroactive films of the polymers are suitable as hole injection/ transport layers for OLEDs. Therefore, the polymers have been tested as hole transporting materials in bilayer OLEDs with Alq_3 as the emitter. The devices containing HTL of the synthesized materials exhibited turn-on voltages of 5.2 – 6.2 V, photometric efficiencies of about 1.6 – 2.5 cd/A and a maximum brightness of 300 – 1510 cd/m^2 at 23 V.

Acknowledgements. The OLED materials were develop in the frame of project funded by a grant No. S-MIP-17-64 from the Research Council of Lithuania.

References

1. S. Lengvinaite et al., Synth. Met. **158**, 213–218 (2008).
2. B. Souharce et al., Macromolecular Rapid Communications. **30**, 2158–1262 (2009).
3. Y. Tao et al., Chem Soc Rev. **10**, 2943–2970 (2011).

INFLUENCE OF ARM STRUCTURE ON THE SELF-ORGANIZATION OF EIGHT-ARMS STAR-SHAPED POLY-2-ALKYL-2-OXAZOLINES IN AQUEOUS SOLUTIONS

T. Kirila, A. Smirnova, A. Blokhin, A. Razina, A. Tenkovtsev, A. Filippov

Institute of Macromolecular Compounds of Russian Academy of Sciences

The aim present work was the study of solution behavior of new eight-arms star-shaped polyoxazolines, which are characterized by arm grafting to upper rim of calix[8]arene. The arms were the block-copolymer polyisopropyloxazoline-b-polyethyloxasoline (PiPOZ-b-PeOZ), block-copolymer polyethyloxasoline-b-polyisopropyloxazoline (PeOZ-b-PiPOZ) and the gradient copolymer of ethyloxazoline and isopropyloxazoline (eOZ-grad-iPOZ). The component ratio was 1:1 for all samples.

Molar masses (M_w) and hydrodynamic characteristics of synthesized samples was determined by the methods of molecular hydrodynamics and optics.

Samples	M_w, Da	R_h, nm	N_a	L_a, nm	[η], cm^3/g
PeOZ-b-PiPOZ	10000	2.3	10	3.7	4.92
PiPOZ-b-PeOZ	12100	3.0	12	4.7	4.43
eOZ-grad-iPOZ	14600	2.9	14	5.5	7.73

The polymers PiPOZ-b-PeOZ, PeOZ-b-PiPOZ and eOZ-grad-iPOZ have similar M_w. Consequently, the degree of polymerization N_a and length of the oxazoline chains for different stars are close. The hydrodynamic radii R_h are significantly less than arm contour length L_a of arms. Therefore, the arms are folded. At the same time, low values of the intrinsic viscosity of the solutions indicate high intramolecular density of star-shaped poly-2-alkyl-2-oxazolines.

Aqueous solutions of samples within concentration range from 0.027 to 0.0015 g/cm^3 were studied using the static and dynamic light scattering and the turbidimetry. The values and temperature dependence of the light scattering intensity, optical transmission, hydrodynamic dimensions and fraction of particles present in the solutions were obtained.

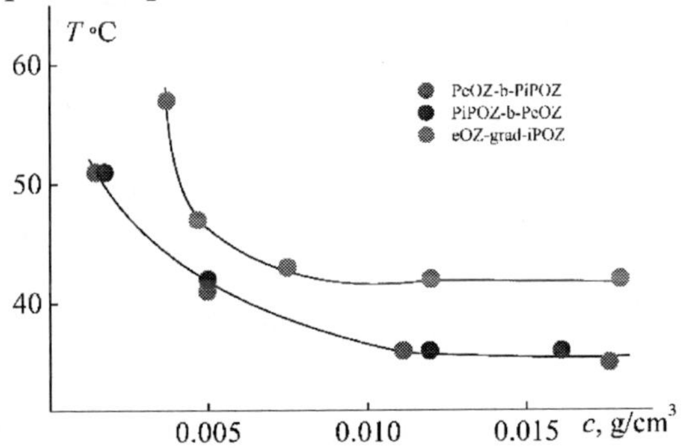

It was shown, that phase separation temperature T_{pt} for eOZ-grad-iPOZ is higher than T_{pt} for PiPOZ-b-PeOZ and PeOZ-b-PiPOZ. Besides, T_{pt} are low than T_{pt} for similar stars with arm grafting to lower rim of calix[8]arene [1].

Acknowledgements. The financial support was provided by Russian Foundation for Basic Research (project No 18-33-00153)

References

1. T.U. Kirila et al., Polym.Sci. Ser. A, **59**, 826–838 (2017).

SYNTHESIS AND PROPERTIES OF TRIPHENYLETHYLENE DERIVATIVES CONTAINING CARBAZOLE AND ACRIDAN MOIETIES

R. S. Bernard, G. Sych, J. V. Gražulevičius

Department of Polymer Chemistry and Technology, Kaunas University of Technology, Kaunas, Lithuania, ronit.bernard@ktu.edu

Organic electroactive compounds gained worldwide interest in the search for alternative materials for large area, flexible, lightweight, and energy efficient optoelectronic devices. In the following work, synthesis and investigation of thermal, optical properties of triphenylethylene derivatives based on carbazole and 9,9-dimethylacridine were synthesized [1]. Absorption and emission spectra of the dilute solutions and the films of compounds are shown in Fig. 1. Compounds 1 and 2emitted greenish light in solid state with the intensity maxima at 448nm and 471 nm and PLQY of 17% and 29% respectively [2]. Aggregations induced enhanced emission properties of the synthesized luminogens were investigated. Thermal properties of the compounds were estimated by DSC and TGA analysis, which confirmed the thermal stability of the target molecules and high morphological stability of their glasses. The investigated compounds showed glass formation abilities and high decomposition temperatures. OLED's were fabricated utilizing the synthesized compounds as emitting materials. The devices containing the layers of two derivatives of acridan and carbazole as emitting layers were fabricated. The highest brightness of 5700 cd/m^2 and power efficiency of 3.2 lm/W was achieved for 9,9-dimethylacridine derivative [3].

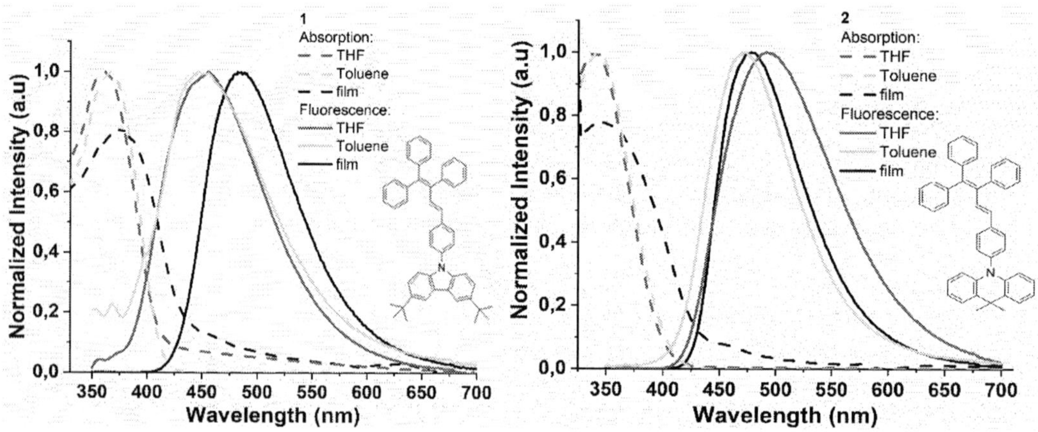

Fig. 1. UV-VIS absorption and fluorescence spectra of toluene, THF solutions and thin films of derivatives of triphenylethylene

References

1. C. W. Tang et al., Organic Electroluminescent Diodes, Appl. Phys. Lett, 51,913-915 (1987).
2. C. Adachi et al., Appl. Phys. Lett, 101, 093306 (2012).
3. D. Gudeika et al., RSC adv, 6, 2191-2201 (2016).

BICARBAZOLE-BASED POLYMERIC NETWORK AS A MIXED HOST FOR EFFICIENT SOLUTION-PROCESSED RED ORGANIC LIGHT EMITTING DIODES

D. Blazevicius[1], D. Tavgeniene[1], S. Grigalevicius[1], D. K. Dubey[2], M. Singh[2], S. Sahoo[2], J. H. Jou[2]

[1] Department of Polymer Chemistry and Technology, Kaunas University of Technology, Radvilen plentas 19, LT 50254, Kaunas, Lithuania, dovydas.blazevicius@ktu.lt
[2] Department of Materials Science and Engineering, National Tsing Hua University, Hsinchu 30013, Taiwan

Fabrication of multilayer organic light-emitting diodes through the solution process presents challenges, especially regarding dissolution of the first layer during deposition of a second layer. Here, we have demonstrated highly efficient solution processed red organic light emitting diodes utilizing a novel electron confining cross linkable small molecules of 9,9′-bis(4-vinylphenylmethylen)[3,3′]bicarbazole (**3**) as a hole transporting and co-host material. Synthesis of the material **3** was carried out according to synthetic route shown in Scheme 1.

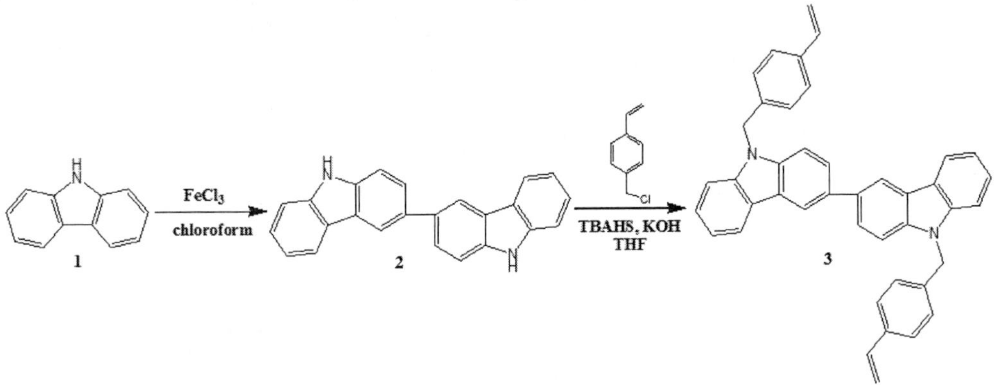

Scheme 1

The measured energy level suggests that compound **3** facilitates the injection of the hole and effectively blocks electron to realize high efficiency, especially at high luminance. Compared with the corresponding single host device, the luminance of the mixed host device was enhanced from 8,351 cd·m^{-2} to 15,200 cd·m^{-2}, an increment of 82%; power efficiency and current efficiency also show an increment of 25% and 20 % at 100 cd m^{-2}, respectively. The use of crosslinked material improved film integrity and hence device brightness. The enhanced charge carrier balance and broadened exciton recombination zone due to the mixed host contribute to the improvement of device performance.

Acknowledgements. The OLED materials were developed in the frame of project funded by a grant No. S-MIP-17-64 from the Research Council of Lithuania.

References

1. Jou J. H., Li T. H., Kumar S., An C. C., Agrawal A., Chen S. Z., Fang P. H., Krucaite G., Grigalevicius S., Grazulevicius J. V., and Sung C. F. Enabling high-efficiency organic light-emitting diodes with a cross-linkable electron confining hole transporting material // Org. Electron. – 2015. – Vol. 24 – P. 254–262.
2. Jou J. H., Kumar S., Agrawal A., Li T. H., and Sahoo S. Approaches for fabricating high efficiency organic light emitting diodes // J Mater. Chem C. – 2015. – Vol. 3 – P. 2974–3002.

TRI AND TETRAPHENYLETHENYL, SUBSTITUTED CARBAZOLE DERIVATIVES: SYNTHESIS, CHARACTERIZATION AND EXHIBITING AIEE AS EFFICIENT HOLE-TRANSPORTING OLED EMITTERS

S. Nasiri, M. Cekaviciute, J. Simokaitiene, A. Petrauskaite, D. Volyniuk, V. Andruleviciene, O. Bezvikonnyi, J. V. Grazulevicius

Department of Polymer Chemistry and Technology, Kaunas University of Technology, Kaunas, Lithuania

New efficient carbazole-based emitters containing tetra-/triphenylethene units were developed for boosting efficiency of non-doped fluorescent organic light-emitting diodes. Comparable study of the properties of derivatives containing one or two tetra-/triphenylethenyl units was performed using various experimental and theoretical techniques. Depending on the substitution pattern, emitters exhibited strong blue or green emission, which was enhanced by aggregation. Compounds with two substituents showed higher glass transition temperatures (up to 120 °C) and lower ionization potentials (of ca. 5.15 eV) comparing to mono substituted derivatives. Time-of-flight hole drift mobility values of the studied compounds reached 10^{-2} cm^2/Vs at high electric fields. Non-doped fluorescent OLEDs based on carbazole derivative containing two tetraphenylethenyl units demonstrated extremely high external quantum efficiency as for simple fluorescent organic light-emitting devices, which reached 5.32 %.

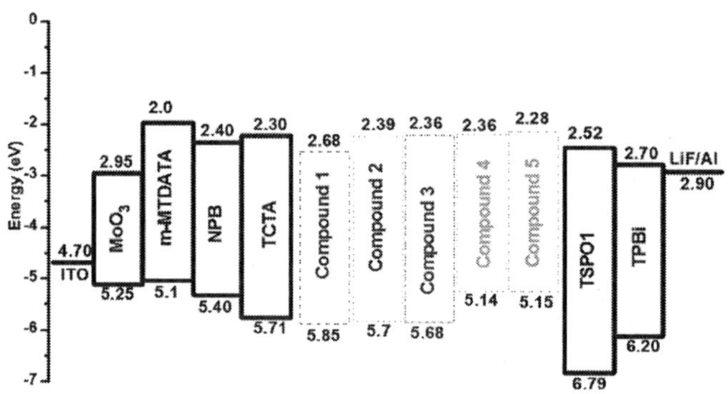

Fig. 1. The equilibrium energy diagrams for the studied devices

Acknowledgements. This research was funded by the European Regional Development Fund according to the supported activity 'Research Projects Implemented by World-class Researcher Groups' under Measure No. 01.2.2-LMT-K-718. Nataliya Kostiv is acknowledged for the initial tests of photophysical properties of the compounds.

References

1. J. Mei, N.L.C. Leung, R.T.K. Kwok, J.W.Y. Lam, B.Z. Tang. Aggregation-Induced Emission: Together We Shine, United We Soar! Chem. Rev., 115 (21), 11718–11940 (2015).
2. X. Shi, C. Yu, H. Su, R. Kwok, M. Jiang, Z. He, J. Lam and B. Tang, A red-emissive antibody–AIE conjugate for turn-on and wash-free imaging of specific cancer cells. Chem. Sci., 8, 7014-7024 (2017).

WHITE SOLUTION-PROCESSED OLEDS BASED ON EMITTERS WITH EITHER PHOSPHORESCENCE, PROMPT OR THERMALLY ACTIVATED DELAYED FLUORESCENCE

D. Volyniuk[1], K. Leitonas[1], J. Simokaitiene[1], E. Skuodis[1], M.D. Thiyagarajan[2], U.M. Balijapalli[2], M. Pathak[2], K. Sathiyanarayanan[2], P. Arsenyan[3], J.V. Grazulevicius[1]

[1] *Kaunas University of Technology, Department of Polymer Chemistry and Technology, Radvilenu pl. 19, LT-50254, Kaunas, Lithuania; e-mail: dvolynyuk@gmail.com*
[2] *Chemistry Department, School of Advanced Sciences, VIT University, Vellore, 632014, Tamil Nadu, India.*
[3] *Latvian Institute of Organic Synthesis, Aizkraukles 21, LV-1006, Riga, Latvia;*

Due to high conversion efficiency of electrical power to visible emission, high colour quality, light weight, flexibility and low operating voltage of organic light emitting diodes (OLEDs), they are applied in flat panel displays and modern lighting equipment [1]. Although, a great number of OLEDs was reported, there is still a challenge to improve colour purity of white OLEDs for lighting technologies. Aiming to develop solution-processed OLEDs with high quality of white electroluminescence, blue fluorescent emitter N,N′-di(1-naphthyl)-N,N′-diphenyl-(1,1′-biphenyl)-4,4′-diamine (NPB) [2], green emitter 4,6-di(9,9-dimethylacridan-10-yl)isophthalonitrile (DAcIPN) exhibiting thermally activated delay fluorescence (TADF) [3], and three phosphorescent iridium (III) complexes with orange or deep-red phosphorescence [4] were selected to use in this work. Light-emitting layers of white hybrid OLEDs were spin-coated from solutions containing above mentioned three different blue, green and orange(red) emitters.

Using orange phosphorescent iridium (III) complex white hybrid OLEDs with human-eyes-friendly electroluminescence were fabricated. The best device was characterized by electroluminescence with CIE1931 coordinates of (0.335, 0.392), colour temperature of 2910K and colour rendering index of 72 and maximum external quantum efficiency of 8.7 %.

Using deep-red emitter phosphorescent iridium (III) complexes hybrid solution-processed OLEDs were obtained. They exhibited high-quality white electroluminescence with colour rendering index reaching of 85. The best devices were characterized by maximum brightness exceeding 10000 cd/m^2 and high external quantum efficiency of 6.26 % [4].

High efficiency of these white hybrid OLEDs is explained by harvesting of triplets through TADF and phosphorescence generated under electrical excitation. The blue fluorescent emitter additionally acted as the host for green and red emitters in the light-emitting layer. Ultralow concentrations of these low-energy emitters were used, thus ensuring full energy transfer from the host to green and orange (red) phosphorescent iridium (III) complexes.

Acknowledgements. This work was supported by the project of scientific co-operation program between Latvia, Lithuania and Taiwan (grant No. S-LLT-19-4).

References

1. S Reineke, et al., Nature **459 (7244)**, 234–238 (2009).
2. J. Liu, Appl. Phys. A, **123 (191)**, 1–6 (2017).
3. E. Skuodis, et al., Org. Electron., **63,** 29–40 (2018).
4. P. Arsenyan, et al., Inorg. Chem., **58 (15)**, 10174–10183 (2019).

THIOL-EPOXY LINSEED OIL-BASED POLYMERS

S. Kasetaite[1], J. Ostrauskaite[1], A. Serra[2]

[1] *Department of Polymer Chemistry and Technology, Kaunas University of Technology, Kaunas, Lithuania,* *sigita.kasetaite@ktu.lt*
[2] *Department of Analytical and Organic Chemistry, University Rovira i Virgili, Tarragona, Spain*

Bisphenol A diglycidyl ether is the most commonly used epoxide in thiol-epoxy click reactions. However, bisphenol A is considered to cause endocrine diseases [1] and pollutes water [2]. Therefore, researchers are interested in the replacement of bisphenol A diglycidyl ether with epoxides obtained from renewable resources. One of the options is natural oils as they are cheap, available, and easily modifiable [3]. Therefore, epoxidized linseed oil (ELO) was chosen for thermal thiol-epoxy reactions with benzene-1,3-dithiol (1,3BDT) and pentaerytritol tetrakis (3-mercaptopropionate) (PETMP). Curing formulations were prepared by mixing stoichiometric amounts of epoxy/thiol groups (1:1). 1-Methylimidazole (1MI) was used as catalyst. The DSC thermograms corresponding to thermal curing of ELO/1,3BDT and ELO/PETMP stoichiometric formulations with 1-5 phr of 1MI are shown in Fig. 1.

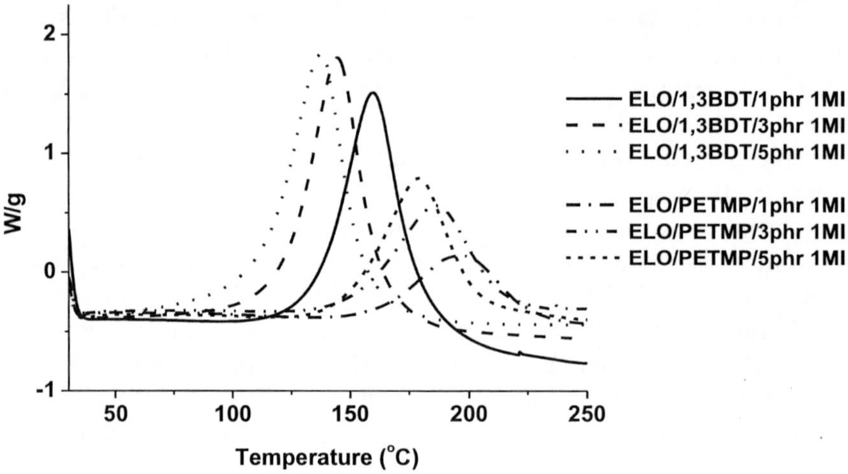

Fig. 1. Calorimetric curves corresponding curing of ELO with different thiols using 1MI as catalyst.

1,3BDT is the most reactive thiol with the lowest activation temperature due to the higher nucleophilicity of the thiophenolate anion compared to PETMP. The reaction of ELO with PETMP starts at about 150 °C and finishes at high temperature which indicates that the reactivity is rather low. The biggest amount of heat (443.8 J/g) was released during the curing of ELO with 1,3BDT using 5 phr of 1MI. According to these results, 5 phr of 1MI is recommended in the compositions of ELO/1,3BDT and ELO/PETMP and the curing process should be carried out at the following schedule: 125, 150, 200, 225 °C for 1 h at each temperature. Thermal and mechanical characteristics of the polymers are of the main importance in their applications. Thus, thermomechanical and thermal stability studies of resulting polymers ELO/1,3BDT and ELO/PETMP with 5 phr of 1MI have been carried out.

Acknowledgements. Financial support from the EU ERDF, through the INTERREG BSR Programme, (ECOLABNET project #R077) is gratefully acknowledged.

References

1. J. Michalowicz, Environ. Toxicol. Pharmacol., **37**, 738–758 (2014).
2. Y. Onundi et al., Green Chem., **19**, 4234–4262 (2017).
3. C. Zhang et al., Prog. Polym. Sci., **71**, 91–143 (2017).

RAFT (CO)POLYMERIZATION OF CARBAZOLE-CONTAINING STYRENE MONOMERS OF ELECTRON-DONOR AND ELECTRON-ACCEPTOR TYPES

A. A. Vaitusionak[1,2], I. V. Vasilenko[1], S. V. Kostjuk[1,2], G. Sych[3], A. Tomkeviciene[3], J. V. Grazulevicius[3]

[1] *Department of Chemistry, Belarusian State University, Minsk, Belarus, vaitusionak@bsu.by*
[2] *Research Institute for Physical Chemical Problems of the Belarusian State University, Minsk, Belarus*
[3] *Department of Polymer Chemistry and Technology, Kaunas University of Technology, Kaunas, Lithuania*

Thermally activated delayed fluorescent (TADF) materials have received increasing attention as effective emitters for producing highly efficient and low-cost organic light-emitting diodes (OLEDs) since they exhibit various advantages such as heavy-metal-free structures and 100% theoretical internal quantum efficiency [1]. TADF properties have been transferred from small molecules [2] to polymers [3], however, up to now, only a few studies on TADF polymer emitters have been reported.

Herein we report the reversible addition-fragmentation chain-transfer (RAFT) (co)polymerization of carbazole-containing styrene derivatives, 9-(4-vinylphenyl)carbazole (M1) and 9-(2,3,5,6-tetrafluoro-4-vinylphenyl)carbazole (M2) (Scheme 1) initiated by 2-(dodecylthiocarbonothioylthio)-2-methylpropionic acid (DDAT)/azobisisobutyronitrile (AIBN) system in 1,4-dioxane (DO) at 70 °C for the first time.

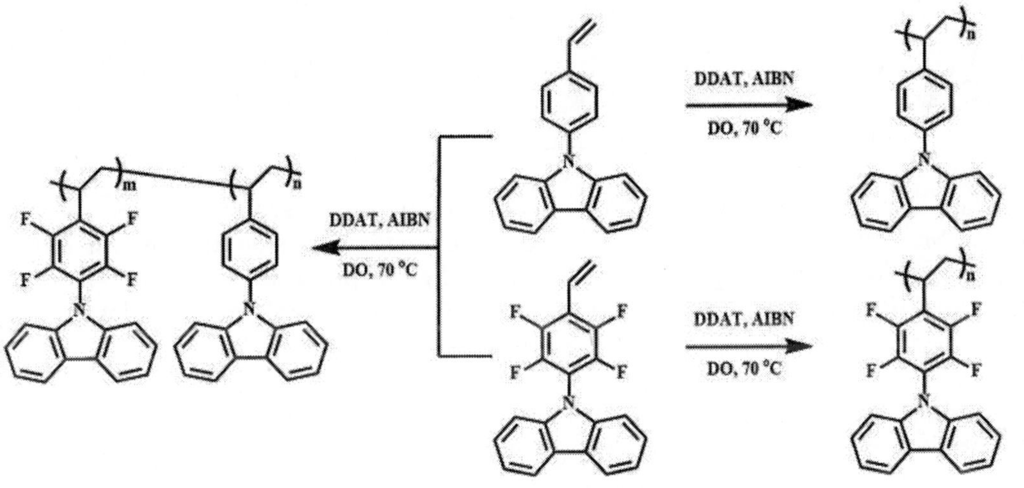

Fig. 1. Scheme of RAFT (co)polymerization of carbazole-containing styrene derivatives.

Polymerization of both monomers proceeds smoothly in a controlled fashion to afford polymers with close to theoretical values of molecular weight. In addition, polymers were characterized by narrow molecular weight distribution (M_w/M_n=1.16). The obtained (co)polymers were also characterized via NMR spectroscopy, size exclusion chromatography, differential scanning calorimetry, UV and fluorescence spectroscopies.

Acknowledgements. This work was supported by European Union's Horizon 2020 Research and Innovation Programme under the Marie Skłodowska-Curie grant agreement No 823720.

References

1. Q. Wei et al., Macromol. Rapid Commun., **40**, 1800570-1800588 (2019).
2. Q. Wei et al., Adv. Opt. Mater., **6**, 1800512-1800542 (2018).
3. K. Albrecht et al., Angew. Chem., **54**, 5677-5682 (2015).

THE STUDY OF SPHERICAL AND CYLINDRICAL BRUSHES WITH THERMOSENSITIVE POLYISOPROPYLOXAZOLINE CHAINS

S. Rodchenko[1], **A. Amirova**[1], **S. Milenin**[2], **M. Kurlykin**[1], **A. Tenkovtsev**[1], **A. Filippov**[1]

[1] *Institute of Macromolecular Compounds of the Russian Academy of Sciences, Saint-Petersburg, Russia, srfm.rodchenko@gmail.com*
[2] *Enikolopov Institute of Synthetic Polymeric Materials of the Russian Academy of Sciences, Moscow, Russia*

The aim of this work was to investigate the macromolecule self-organization of spherical and cylindrical brushes with thermosensitive polyisopropyloxazoline chains in aqueous solutions. Thermosensitive spherical brushes with four ($PiPrOx_4$) and eight ($PiPrOx_8$) arms with carbosilane dendrimer as a core, and cylindrical brush with polydimethylsiloxane backbone and polyisopropyloxazoline side chains (PDMS-*graft*-PiPrOx) were studied using light scattering and turbidimetry methods.

Sample	*M, g/mol*	*c, g/cm*^{3}	T_1, ℃	T_2, ℃
$PiPrOx_4$	14800	0.0054	44	46
$PiPrOx_8$	31800	0.0044	39	43
PDMS-*graft*-PiPrOx	400000	0.0049	31	36

Fig. 1. Phase separation temperatures of the studied samples.

At certain concentration, the highest temperatures of phase separation were obtained for spherical brush $PiPrOx_4$. The latter can be caused by better core shielding in four-arm sample. On the other hand, PDMS-graft-PiPrOx had the lowest phase transition temperatures. The architecture of macromolecules and a larger molecular weight are the reasons for these phenomena.

Acknowledgements. This work has been supported by the State Program for Support of Leading Scientific Schools (grant no. 14.W03.31.0022).

DIMETHOXY-SUBSTITUTED TRIPHENYLAMINE DYES FOR SOLID STATE DYE SENSITIZED SOLAR CELLS

D. Gudeika[1], K. Dabrovolskas[1], S. Benhattab[2], M. B. Manaa[2], N. Berton[2], J. Bouclé[3], F. Tran Van[2], B. Schmaltz[2], D. Volyniuk[1], J. V. Grazulevicius[1]

[1] *Department of Polymer Chemistry and Technology, Kaunas University of Technology, Kaunas, Lithuania, gudeika.dalius@gmail.com*
[2] *Laboratoire de Physico-Chimie des Matériaux et des Electrolytes pour l'Energie (PCM2E), Université de Tours, Tours, France*
[3] *Institut XLIM UMR 7252, Université de Limoges/CNRS, Limoges, France*

In recent years, the investigation on solar cells using organic semiconducting derivatives has made considerable advance [1]. Among the promising photovoltaic technologies, dye-sensitized solar cells (DSSCs) have received much attention recently due to their low cost and relatively high efficiency [2]. Most conventional organic dyes possess the push-pull structure of "donor-conjugated bridge-acceptor". Typically, organic dyes used for efficient solar cells are required to afford intense and broad absorption in the visible spectral region. Aside from donor units, the electron acceptor plays a significant role in the performance control of DSSCs [3]. Introducing an additional electron-donating moiety into derivative of triphenylamine can enhance charge separation, which result in high overall conversion efficiency [4]. Due to the certain peculiarities of the structure of triphenylamine, electron acceptor groups can be introduced. They can affect the molar extinction coefficients, absorption spectra, and the HOMO, LUMO energy levels of the derivatives. Rhodanine 3-acetic acid and cyanoacetic acid were introduced into triphenylamine-based dyes as acceptor moieties [5].

In this paper, easily synthesized dyes based on dimethoxy-substituted triphenylamines were obtained by a *Knoevenagel* condensation. Two different dyes containing cyanoacrylic acid and rhodanine-3-acetic acid moieties as an electron acceptors were prepared. These dyes show interesting electronic and optoelectronic properties and, as we show, constitute potential candidates to be used in solid state dye sensitized solar cells based using the structure of FTO/TiO_2/Dye/Spiro-OMeTAD/Ag. In addition, we studied Kerr and two-photon absorption effects of the synthesized derivatives to understand how different acceptor groups influence NLO properties.

Acknowledgement. This research was funded by a grant (No. S-LZ-19-2) from the Research Council of Lithuania.

References
1. F. J. Kahle, et al., Adv. Energy Mater., **7**, 1700306 (2017).
2. M. Grätzel, et al., Nature, **353,** 737(1991).
3. W. Zeng, et al., Chem. Mater., **22,** 1915 (2010).
4. L. Zhang, et al., Tetrahedron, **66,** 3318 (2010).
5. X. Jiang, et al., Adv. Funct. Mater., **21,** 2944 (2011).

NON-DOPED ORGANIC LIGHT-EMITTING DIODES BASED ON COMPOUNDS EXHIBITING AGGREGATION INDUCED EMISSION ENHANCMENT

U. Tsiko, G. Sych, O. Bezvikonnyi, J. Simokaitiene, D. Volynuik, J. V. Grazulevicius

Department of Polymer Chemistry and Technology, Kaunas University of Technology, Kaunas, Lithuania, uliana.tsiko@ktu.edu

Organic semiconductors have been broadly studied for numerous applications including organic light-emitting diodes, organic field-effect transistors and organic solar cells [1]. The luminogenic building blocks, such as phenylethylene and other rotors can significantly change the solid-state emission behaviour from the aggregation caused quenching to aggregation induced emission, which is a useful technique for fabrication of the low-cost, bright and efficient organic light emitting diodes [2].

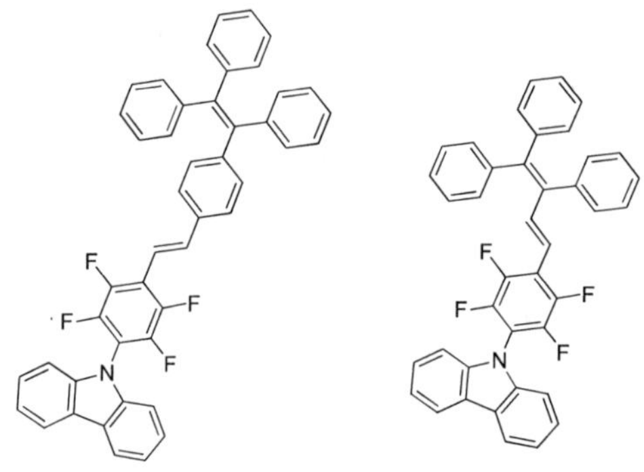

C4FS1 C4FS2

The synthesized compounds C4FS1 and C4FS2 were characterized as materials exhibiting both aggregation-induced emission enhancement and bipolar charge-transporting properties. The compound having tetraphenylethylene moiety demonstrated blue-greenish emission with maximum fluorescence intensity wavelengths of 511 nm for solid-state sample and 498 nm for toluene solution, whereas the compound having triphenylethylene moiety showed emission maxima at 502 nm for solid film and 429 nm for toluene solution. The compounds demonstrated high photoluminescence quantum yields in solid state of 57% and 27% respectively. Compounds C4FS1 and C4FS2 showed high thermal stability with 5% weight loss temperatures of 331°C and 361°C respectively and high glass-transition temperatures of 170 and 211°C correspondingly.

Non-doped organic light emitting diodes based on C4FS1 and C4FS2 as emitting materials were fabricated and exhibited green electroluminescence. The best non-doped device containing C4FS1 as emitter showed turn-on voltage of 9.1 V, maximum brightness of 11 800 cd/m^2, maximum current efficiency of 4.5 cd/A, and EQE of ca. 1.7%.

References

1. Y. Hong et al., Chem. Soc. Rev., **40**, 5361–5388 (2011).
2. J. Mei et.al., Chem. Rev., **115**, 11718–11940 (2015).

KINETICS OF VEGETABLE OIL EPOXIDIZED ACRYLATE PHOTOINDUCED CURING

A. Barkane[1], S. Gaidukovs[1], O. Platnieks[1], M. Jurinovs[1], Y. Habibi[2], N. Kasmi[2]

[1] *Institute of Polymer Materials, Faculty of Materials Science and Applied Chemistry, Riga Technical University, P. Valdena 3/7, LV-1048 Riga, Latvia, Anda.Barkane@rtu.lv*
[2] *Luxembourg Institute of Science and Technology, Department of Materials Research and Technology (MRT), Esch-sur-Alzette, Luxembourg*

Based on the recently received promising results on plant oil-based epoxy acrylate (EA) as an important type of free-radical UV-curable bio-based prepolymers, reassuring the merits of facile synthesis, less pollution, low energy consumption[1,2,3], this research was realized for additive manufacturing applications purposes. Bio-based prepolymer formulations were based on soybean oil epoxidized acrylate (AESO) oligomer that was blended with specific functional acrylic monomers to increase flexibility, water resistance[4], tensile strength and adhesion properties[5]. System was polymerized by using various concentrations of 2,4,6-trimethylbenzoyldiphenylphosphine oxide (TPO) photoinitiator with absorption band at 379 nm and two different types of UV-light source: LED and laser. Systems crosslinking kinetics, depending on TPO concentration and UV-light source was investigated by Fourier Transformation Infrared spectrometry (FTIR) and RAMAN spectroscopy. Crosslinking network characteristics and gel fraction were also obtained.

Acknowledgements. This research is funded by the M-era.net 2017 project "3D Printable Innovative Biobased Materials for Wood Mimics", 3DPrintInn; Nr.1.1.1.5/ERANET/18/05

References

1. Q. Wu et al., ACS Sustain. Chem. Eng., **6**, 8340 (2018).
2. H. Pelletier et al., Appl. Polym. Sci., **99**, 3218 (2006).
3. S. Rengasamy et al., Prog. Org. Coatings, **76**, 78 (2013).
4. S. A. Madbouly et. al., Elsevier Inc., 99-125 (2016).
5. M. Khalid et. al. Eng. Sci. Technol. Rev., **3**, 153-162 (2008).

CHARACTERIZATION OF BITUMEN MASTICS PROPERTIES

A. Ābele[1], R. Meri Merijs[1], R Bērziņa[1], J Zicāns[1], V Haritonovs[2]

[1] *Institute of Polymer Materials, Faculty of Materials Science and Applied Chemistry, Riga Technical University, Riga, Latvia, Agnese.Abele_1@rtu.lv*
[2] *Department of Roads and Bridges, Faculty of Civil Engineering, Riga Technical University, Riga, Latvia*

The asphalt pavement has been exposed to environmental factors and high loads that leads to deformation and decline durability of asphalt pavement. Despite the fact that binder comprises only approximately 5% to form asphalt pavement, it play an important role in asphalt pavement service life [1]. Firstly, bitumen chemical content - saturates (S), aromatics (A), resins (R) and apshalthenes (A) – depend on crude oil and manufacturing process. In result it determines whether bitumen can be useful for road pavement, because SARA fractions provide adhesion, viscosity and stiffness [2]. Secondly, in exploitation time deterioration of properties of binder can be observed over time by environmental factors. Bitumen become more polar, stiffer and changes rheological behavior [3]. To prolong binder service life, as well as maximally to increase re-use of secondary resources, scientists perform binder modification. Modification can be realized in several ways. Thus one can modify bitumen with thermoplastic polymer (i.e. polyethylene, polypropylene, ethylene vinyl acetate, styrene-butadiene rubber, styrene-ethylene/butylene-styrene [4]) or filler (i.e. fly ash, stone dust, cement) [5].

In this current work we investigated several industrial by-products (red mud, by-pass dust and fly ash) influence on elastomer modified bituminous material rheological and structural parameters.

By-products were added (6, 7, 8 wt. %) in 70/100 grade bitumen modified with 4 wt. % styrene-butadiene-styrene copolymer (SBS). Rheological properties (viscosity, rutting resistance) are determined by dynamic shear rheometer. Results show that addition of SBS considerably increases viscosity (by approx. 300%) due to better molecular interaction between the matrix and SBS. At the same time bituminous materials with by-products viscosities values are higher than neat bitumen or elastomer modified bitumen. The same tendency is observed at $G^*/\sin\delta=1$kPa corresponding temperatures ($T_{critical}$), which characterizes rutting resistance. $T_{critical}$ increase not only with addition of SBS but also with addition of by-products; the highest $T_{critical}$ is reached at lower (6 wt.%) filler amount. This means better interaction between bitumen, SBS and the filler. Higher filler contents increase stiffness of the material, which can lead to several problems during laying and exploitation (e.g. cracking).

Rheological behavior of bituminous mastics has been correlated to structural changes. Interaction between the bitumen SARA fractions, SBS and by-products has been observed by differential scanning calorimeter (MDSC). These interactions lead to the shift of glass transition temperatures to lower values that mean decreased brittleness.

Acknowledgements. This work has been financially supported by the Applied Research Project "Innovative use of industrial byproducts for sustainable asphalt pavement mixture" No Z18/1-0237 of the Latvian Council of Sciences

References

1. http://www.interchem.at [29.05.2019]
2. M. Paliukaitė et al. , The 9th Conference Environmental Engineering. Ar.no:enviro.2014.162 (2014).
3. M.Hamzah et.al., Casp. J. Appl. Sci. Res., 4(6), 15-34 (2015).
4. J.Zhu et.al., Eur Polym J., 54, 18-38 (2014).
5. R.Mistry et.al., Perspect Sci., 8, 307—309(2016).

REAL-TIME PHOTORHEOMETRICAL STUDY OF VANILLIN-BASED RESINS

A. Navaruckienė, J. Ostrauskaitė

Department of Polymer Chemistry and Technology, Kaunas University of Technology, Kaunas, Lithuania

Vanillin-based resins were composed of vanillin diacrylate or vanillin dimethacrylate (VDM) with or without 1,3-benzenedithiol (13BDT) using the diphenyl(2,4,6-trimethylbenzoyl)phosphine oxide (TPOL) as photoinitiator. Dichloromethane (DCM) was used to dissolve vanillin dimethacrylate.

Real-time photorheometry was used to monitor the evolution of thiol-ene and free-radical photocross-linking process. As an example, the dependencies of storage modulus G', loss modulus G", loss factor tanδ, and complex viscosity η* of the resin VDM/13BDT/3TPOL/DCM on irradiation time are presented in Fig. 1. When irradiation of the resin started, the values of storage modulus G', loss modulus G", and complex viscosity η* started to increase indicating the beginning of cross-linking process. The onset of gelation process is described as the gel point (t_{gel}), i.e. the point at which G' and G" modulus curves intersect [1]. As the irradation of the resin proceeded with time, the values of G', G" modulus and η* continued to increase due to the gel aging and settled down into a steady-state indicating the end of the cross-linking process. All vanillin-based resins investigated in this study showed the similar behaviours.

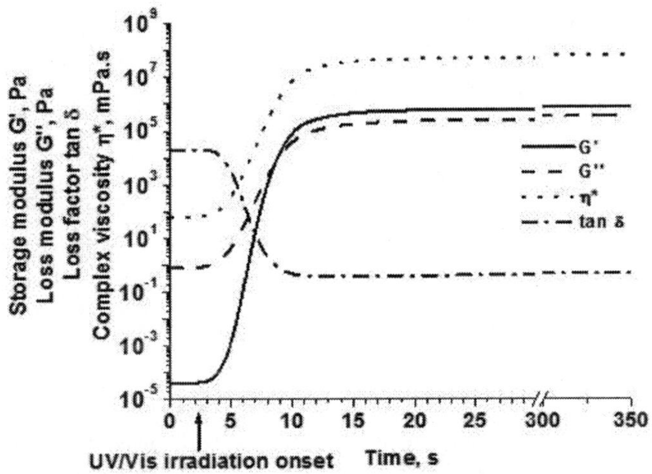

Fig. 1. Dependencies of storage modulus G', loss modulus G", loss factor tan δ, and complex viscosity η* of the resin VDM/13BDT/3TPOL/DCM on the irradiation time

Real-time photorheometrical study showed that vanillin diacrylate is more reactive than vanillin dimethacrylate, as resins with vanillin diacrylate needed longer UV curing time and reached the gel point later. Moreover, it was determined that free-radical photocross-linking of vanillin-based resins is a quicker process than thiol-ene photocross-linking. The influence of DCM and the amount of TPOL on photocross-linking rate and rigidity of the formed cross-linked polymers was investigated as well.

Acknowledgements. This research was funded by the European Social Fund under the measure No. 09.3.3-LMT-K-712 "Development of Competences of Scientists, other Researchers and Students through Practical Research Activities" and the EU ERDF, through the INTERREG BSR Programme, (ECOLABNET project #R077).

References

1. Mezger, T.G. (2011), *The Rheology Handbook, 3rd ed.*, Vincentz Network, Hanover.

REGULATION OF REOLOGICAL PROPERTIES OF FIRE-SAFETY POLYMERIC COMPOSITIONS

R. Iskandarov[1], O. Chulieieva[2], V. Plavan[1], D. Novak[1]

[1] *Kiev National University of Technology and Design, plavan.vp@knutd.edu.ua*
[2] *Private Joint-Stock Company "PLANT PIVDENKABEL", echuleeva@ukr.net*

In the developed fire-safety polyolefin polymer compositions as fillers-antipirens are the use of metal hydroxides, in particular, aluminum and magnesium, magnesite and hydromangesite, which allows to increase fire resistance by absorbing a large amount of heat, neutralize acid gases, reduce the smoke formation [1-3].
One of the problematic issues that arises during the processing of fire-fighting polymer compositions is the increased viscosity. In order to solve this problem, a method for direct regulation of rheological characteristics of polymer compositions has been developed by introducing into their composition a modifier - aminosilane.
Dependence of change of rheological indices from the modifier was investigated by capillary viscometry method [4]. Results of investigations of fire-safe ethylene vinyl acetate copolymer compositions are shown in Fig. 1

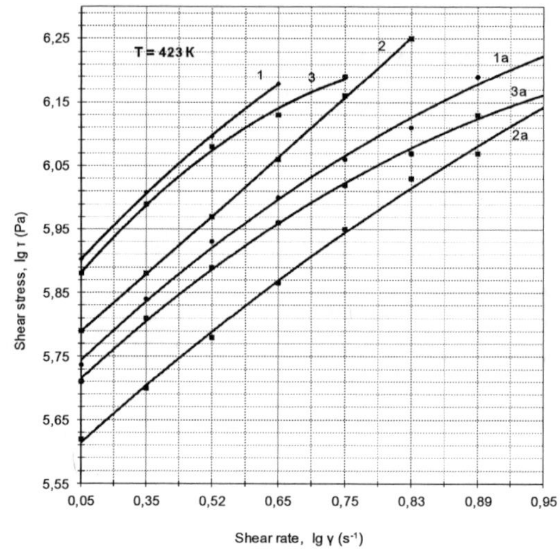

Fig. 1. Dependence of the shear rate on the shear stress of ethylene vinyl acetate copolymer compositions and flame retardants (example 1 - trihydrate of aluminum oxide, 2 - magnesium oxide dihydrate, 3 - hydromagnesite) without modifier and modifier in the amount of 0,6% (Sample 1a -3a)

As a result of the modification of polymeric compositions, the shear stress is reduced from 6.18 Pa to 6.0 Pa for aluminum oxide trihydrate, from 6.06 Pa to 5.86 Pa for magnesium oxide dihydrate, from 6.13 Pa to 5.96 Pa for hydromangeseite while achieving the same shear rate.
Conclusions. The best fire-safety properties of insulations and shells of cable products is provided by a modifier (N- (3-trimethoxysilane) propylbutylamine) in an amount of 0.6%.

References

1. O. Chulieieva, Eastern European Journal of Enterprise Technologies, **5/12**, 65-70 (2017).
2. R. Herbiet, High Performance Filler, **4**, 20 (2005).
3. M. A. Cardenen et al., Polymer Degradation and Stability, **93 (11)**, 2032-2037 (2008).
4. Teplofizicheskiye i reologicheskiye charakteristiki polimerov. Spravochnik pod redaktsiyey akademika YU. S. Lipatova. - K .: Naukova Dumka, 244 (1977).

INVESTIGATION OF UV-CURED ACRYLATED EPOXIDIZED SOYBEAN OIL FILMS WITH DIFFERENT PHOTOINITIATORS

M. Lebedevaitė, **J. Ostrauskaitė**

Department of Polymer Chemistry and Technology, Kaunas University of Technology, Kaunas, Lithuania, jolita.ostrauskaite@ktu.lt

In recent years, the environmental concern of petroleum-derived materials has risen, unfortunately a lot of materials used in film and coating compositions are still not bio-based. Acrylated epoxidized soybean oil (AESO) has become a considerable candidate for replacement of petroleum-derived acrylates in various applications [1]. High functionality and immerse sensitivity to light enabled the wide applications of AESO in acrylic compositions [2].

In this study, AESO was photocross-linked with four different photoinitiators: 2,2-dimethoxy-2-phenylacetophenone (DMPA), diphenyl(2,4,6-trimethylbenzoyl)phosphine oxide (TPO), phenylbis(2,4,6-trimethylbenzoyl)phosphine oxide (BAPO) and ethyl (2,4,6-trimethylbenzoyl) phenylphosphinate (TPOL). 1, 3 and 5 mol.% of mentioned photoinitiators were used in the resins. Chemical structure of the cross-linked polymers was confirmed by IR spectroscopy. The insoluble fraction of the cross-linked polymers was determined by Soxhlet extraction. Mechanical testing of the cross-linked polymer films was performed by tensile test on a BDO-FB0.5TH (Zwick/Roell) testing machine and by compression test with Testometric M500-50CT. Thermogravimetrical analysis was conducted on a Perkin Elmer TGA 4000 instrument. Differential scanning calorimetry analysis was performed on TA Universal DSC Q2000 V24.10 Build 122 instrument.

It was determined that the amount of photoinitiator had the highest influence on the polymer film tensile properties. AESO polymer films with higher amount of photoinitiatior indicated higher Young modulus and tensile strength values, although elongation at failure was reduced. As an example, mechanical characteristics of AESO films prepared with 3 mol. %/ of various photoinitiators are presented in figure 1. All photocross-linked polymers indicated 97-98 % of insoluble fraction and glass transition temperatures above 40 °C.

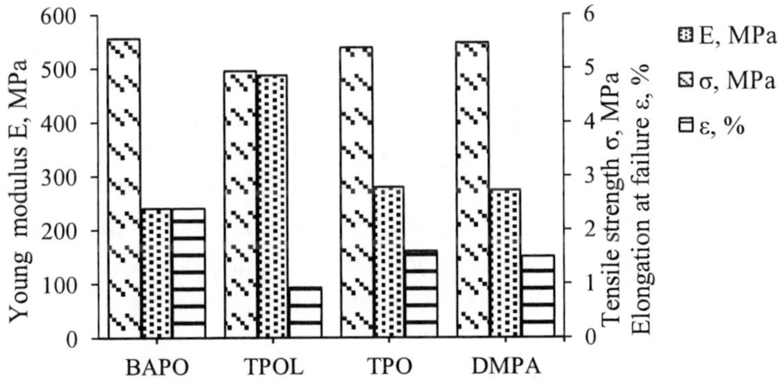

Fig. 1. Mechanical characteristics of AESO films prepared with 3 mol.% of various photoinitiators

Acknowledgements. Financial support from EU ERDF, through the INTERREG BSR Programme (ECOLABNET project #R077) is gratefully acknowledged.

References

1. Liu, Pengfei, et al. Progress in Organic Coatings **134,** 342-348 (2019).
2. Liu, Wendi, et al. Composites Science and Technology **160,** 60-68 (2018).

KNOWLEDGE KIT FOR DESIGN OF NANOMODIFIED POLYOLEFIN MULTILAYER PRODUCTS WITH ENHANCED OPERATIONAL PROPERTIES.

I. Bute, S. Stankevich, O. Starkova, O. Bulderberga, and A. Aniskevich

Institute for Mechanics of Materials, University of Latvia, irina.bute@lu.lv

Despite the rapid development of nanotechnology in the last 20 years, the potential for their use in industrial applications is still far from the expected one. One of the reasons for this is the lack of easy-to-use guidelines and manuals for design of products from nanomodified polymers.

The present work is aimed to develop a scientifically based information platform for manufacturers of polyolefin products and end-users that will guide them in design, production and characterisation of novel nanomodified products, their effective use. This will be realized through the development of a knowledge kit.

The knowledge kit consists of 3 parts: easy-to-use software for analytical calculation of some operational characteristics of multilayer products made from nanomodified polyolefins; a database of properties of nanomodified polyethylene and polypropylene and a guideline.

In the process of working with the calculation module of the knowledge kit, the user must enter data on the multilayer package (number of layers and thickness of each layer) and select from the database material of each layer, type of filler and its percentage. Next, the user needs to select from the proposed list the property that he needs to calculate (density, Young's modulus, coefficient of linear thermal expansion, thermal and electrical conductivity). After the calculations, the module will save all the data as a project that can be reused later.

Formulas for calculating of the properties of a multilayer polyolefin from known models (mixture rule, isostrain assumption by Vougt and Reuss) were proposed and included in the calculation module. The property values of a single layer were entered into the database from the experimental data and/or data sheet. The possibility of adding to the database, as new values of properties, and materials, and fillers was designed. The calculation module provides possibility to save the calculated data as a project for the reuse, as well as their preservation and export to pdf format.

The knowledge package contains data on the type (LDPE, PP, HDPE, MDPE, LLDPE, PE) and name (Eltex TUB 433-NA00, Eltex PF6212AA, Rigidex HD5502S, Rigidex K38-20, 22D730, 23H430, LL6208AF, PLASTICYL™ HDPE1501, PLASTICYL™ PP2001) of materials used and fillers, on the percentage content of the filler (2-15%), on the values of the properties for each type of material and the corresponding filler, as well as data on the source of the obtained data and measurement standards and it was successfully approbated.

It is planned to develop the possibility to predict the calculated values of the properties of both the layer and the multilayer package by interpolation, if the necessary values are not in the database.

The knowledge kit developed can be widely used in industrial applications for design of multilayer products with tailored operational properties.

Acknowledgements. The authors thank European Regional Development Fund project: Development of nano-modified polyolefin multilayer extrusion products with enhanced operational properties (ERDF Project identification No.: 1.1.1.1/16/A/141).

STRUCTURAL RESEARCHES OF POLYETHYLENE COMPOSITIONS FILLED BY COPPER-COATED GRAPHITE AND CARBON NANOTUBES

R. Iskandarov, D. Novak, Y. Budash, V. Plavan

Department of Applied Ecology, Technology of Polymers and Chemical Fibers, Kyiv National University of Technologies and Design, Kyiv, Ukraine, novak.knutd@gmail.com

Studies of carbon filler particles distribution in polymer compositions are important from the standpoint of using such compositions as electrically conductive composite films. This is due to the fact that the process of electrical conductivity occurs only when there is an interconnection of filler particles. According to the results of the particle distribution studies on the film`s surface about its nature and the ability of the film to electroconductivity can be concluded [1, 2].

The prospect of using a mixture of graphite and carbon nanotubes as filler in the manufacture of electrically conductive composites based on polyethylene shown in [3].

The aim of this work is to investigate the structure of polyethylene (PE) films filled with copper-coated graphite (CG) and carbon nanotubes (CNT) in different ratios, to assess the quality of mixing and the uniform distribution of fillers in the polymer composition.

The PE was mixed with the filler in the gap of the heated disk device, which ensured uniform distribution of the filler in the polymer matrix. The mixing quality of the components was determined by the optical method. After mixing the composition mixture was placed in a heated mold. Mold with the compositional mixture is placed in a heat chamber for heating and transition to a highly plastic state, in which the presswork can be made. After staying of materials in the heated chamber for 12-15 min at 160°C, it's compressed at a pressure of 22 MPa using laboratory press [4].

The satisfactory uniformity filler particles distribution was determined for PE compositions filled with a mixture of CG and CNT. The correlation between the uniformity coefficient and the filler content was obtained for PE compositions with 6% of the filler. The uniformity coefficient increases in direct proportion with an increase in the content of CNTs (a decrease in the CG content).

The influence regularities of the ratio of carbon fillers of various nature on their surface concentration in composite films based on PE were established.

A non-linear correlation relation was established between the content of CNTs in the modified additive and the inhomogeneity coefficient of the distribution of filler particles on the surface of the composite film.

Using cluster analysis, the homogeneity of the carbon filler particles distribution on the surface of composite films of various compositions was evaluated.

Developed PE compositions filled with a mixture of CG and CNT (content from 1 to 6 %) can be further used as masterbatches for the production of products for electrical purposes.

References

1. T. Mcnally Polymer-carbon nanotube composites: preparation, properties and applications, Oxford: Woodhead, p. 848 (2011).
2. N.A. Mohd Radzuan, A.B. Sulong, J. Sahari A review of electrical conductivity models for conductive polymer composite, International Journal of Hydrogen Energy, **42(14)**, 9262–9273 (2017). https://doi.org/10.1016/j.ijhydene.2016.03.045J
3. K. Kar Kamal, K. Pandey Jitendra, Rana Sravendra Handbook of Polymer Nanocomposites. Processing, Performance and Application: Volume B: Carbon Nanotube Based Polymer Composites, Springer, Germany, **19**, p. 601 (2015).
4. Y. Budash, D. Novak, V. Plavan Structural and Morphological Characteristics of Polyethylene Composites with Different Conductive Fillers, MaterialePlastice, **53**, no. 4, 693–698 (2016).

POLYMER BRUSH BASED ON ANTHRAZOLINE-CONTAINING DIAMINE

I. A. Valieva[1], M. Y. Goikhman[1,2], I. V. Podeshvo[1], I. V. Gofman[1], R. Yu. Smyslov[1], L. S. Litvinova[1], A. V. Yakimanskii[1,2]

[1]Institute of Macromolecular Compounds of Russian Academy of Sciences, St.Petersburg, Russia
[2]Saint Petersburg State University, Institute of Chemistry, St.Petersburg, Russia
valieva@hq.macro.ru

Polymer brushes are macromolecular objects consisting of molecules tightly grafted to the surface of nanoparticles and thin membranes or to polymer chains. These materials are used in medicine and nanotechnology. We have developed a method for obtaining polymer brushes by forming quaternary bases in the interaction of alkyl iodides with polymers, the elementary unit of which contains fragments of nitrogen-containing heterocycles, for example, anthrazoline. For this purpose, 2,8-bis(4-aminophenyl)pyrido[3,2g]quinoline-4,6-dicarboxylic acid [1] and diethyl 2,8-bis(4-aminophenyl)pyrido[3,2g]quinoline-4,6-dicarboxylate were synthesized.

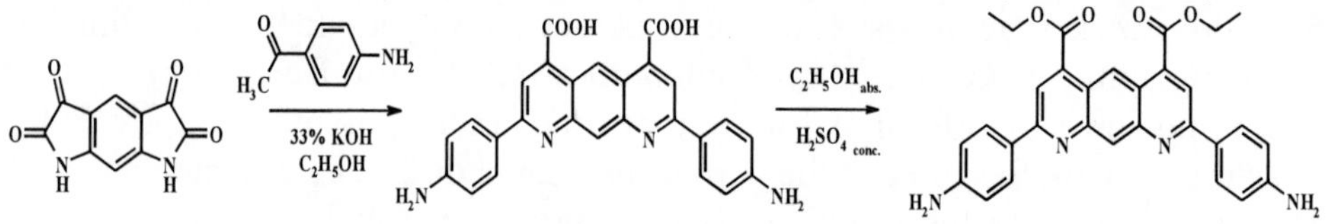

Fig. 1. Scheme of the synthesis of monomer.

Diethyl ether was used as a monomer to obtain a copolyamide containing 30% of antrazoline units. A polymer brush was synthesized from this copolyamide by interaction with decyl iodide [2,3].

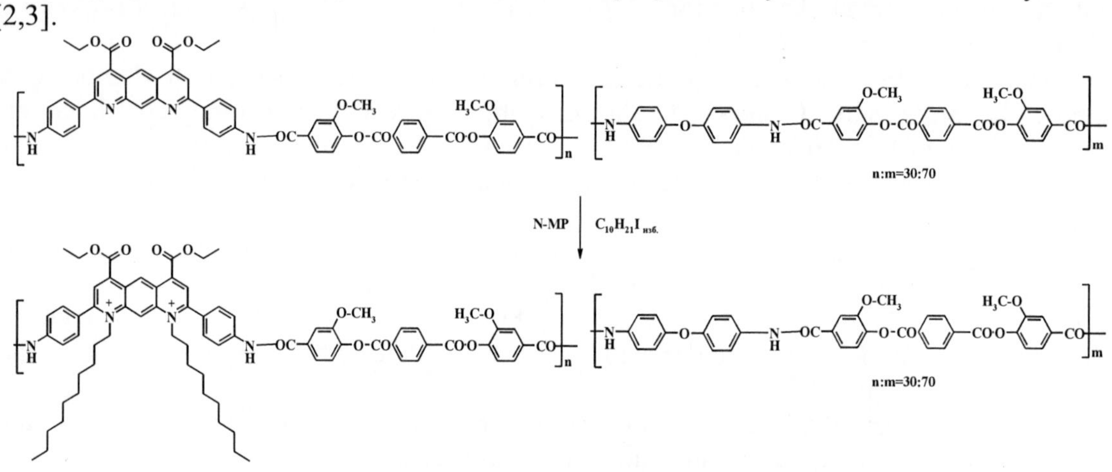

Fig. 2. Scheme of the synthesis of polymer brush.

The structure of the synthesized polymer was proved using the method of ^{1}H-NMR.

Acknowledgements. The work is supported by the project No.14.W03.31.0022.

References

1. M.Ya. Goikhman, I.A. Valieva, I.V. Podeshvo et al., Luminescence, **33**, 559-566 (2018).
2. J.A. Zoltewicz, M.P. Cruskie, Jr., Tetrahedron, **51**(11), 3103-3114 (1995).
3. C.K. Cain et al., Org. Chem., **20**, 4466-474 (1955).

PREPARATION AND CHARACTERIZATION OF α,ω -DIHYDROXY-POLY(DIMETHYLSILOXANE) MODIFIED POLYESTERS FOR TISSUE ENGINEERING

S. Gailiūnaitė[1], T. Kochanė[1], V. Bukelskienė[2], D. Baltriukienė[2], S. Budrienė[1]

[1] *Vilnius University, Faculty of Chemistry and Geosciences, Institute of Chemistry, Naugarduko 24, 03225 Vilnius, Lithuania* sonata.gailiunaite@chgf.stud.vu.lt
[2] *Vilnius University, Life Sciences Center, Institute of Biochemistry, Saulėtekio av. 7, 10257 Vilnius, Lithuania*

Tissue engineering is rapidly developing field, it has become a promising approach to cure a number of damaged tissues or organs [1]. The tissues are grown on a synthetic artificial carcass which is created using UV-curing. The cells will reproduce and functionalize on frame, so the materials chosen for it should be biocompatible, also it must have mechanical strength, be flexible, biodegradable and non-toxic [2]. Poly(dimethylsiloxane) (PDMS) elastomer is one of those materials due to its properties such as: non-toxicity, biocompatibility, thermal stability, chemical and biological inertness. However, its use is limited by hydrophobicity which can cause cellular adhesion on the surface to be short-lived, and it has low mechanical resistance. PDMS could be modified to improve hydrophilicity, and then the ability to use PDMS significantly could be enhanced [3]. One of the possible modification methods is polyesters modification with hydroxyl terminated PDMS.

Polyester films were synthesized from azelaic acid (AA), maleic anhydride (MA), diethylene glycol or triethylene glycol and were chemically modified with hydroxyl terminated PDMS, at various initial molar ratios. Glycidyl methacrylate (GMA)and/or butyl methacrylate (BMA) at various initial molar ratios were used as curing agents to obtain UV-curable films and various size of water soluble salt crystals were used to obtain pores in the films. Chemical structure of films was investigated by FTIR, H^1 NMR and EDS. The films were tested for swelling degree and solubility in hexane, ethanol and water. Films with higher concentration of AA were less soluble in hexane, that's means that PDMS incorporation in polymer was better when AA concentration was higher. The films were less soluble in ethanol when amount of MA was higher, because there is larger number of double bonds in the polyester chain. The amount of Si in cured films varied in the range from 0,6 % to 8,3 % and was higher when films were synthesized with higher amount of AA. Obtained films exhibit good wettability in comparison to commercial PDMS film (Sylgard 184), their water contact angle was lower (76.7-77.6°) than obtained in the Sylgard 184 case (101°). The influence of pore size on properties of films was investigated. The mechanical properties of the films were evaluated by tensile test. Elongation at break (X_R) data suggest that elastic films were obtained using a mixture of GMA and BMA as curing additives and X_R ranges from 120 % to 320 %, which means that X_R was 4-11 times higher than commercial PDMS. The biocompatibility of the formed films was examined by biocompatibility study with rat tooth pulp stem cells. All films were biocompatible.

References

1. M.J. Coenena, et. al., Acta Biomaterialia **79**, 60-82 (2018).
2. Yu et. al., Biomaterials **194**, 1-13 (2019).
3. B.Y. Yoo et. al., Acta Biomaterialia **76**, 56-70 (2018).

pH-RESPONSIVE BEHAVIOR OF ANIONIC POLYMER BRUSHES SYNTHESIZED BY RAFT AND *CLICK* CHEMISTRY REACTIONS

J. Jonikaitė-Švėgždienė, **M. P. Mameniškis, R. Makuška**

Institute of Chemistry, Vilnius University, Vilnius, Lithuania, jurate.jonikaite-svegzdiene@chf.vu.lt

An increasing need for "smartness" in biomedical and engineering materials has gain a growing interest for synthetic polymers that exhibit environmentally responsive behavior [1]. pH-responsive polymers are a group of stimuli-responsive polymers that can respond to solution pH by undergoing structural and property changes such as chain conformation, solubility, or configuration. These unique properties make pH-responsive polymers useful in various applications such as drug/gene delivery, surfaces, sensors, membranes, and chromatography [2-3].

In this study, polymer brushes were synthesized by "grafting to" strategy, combining RAFT polymerization and two *click* chemistry reactions, namely, epoxy-thiol and azide-alkyne. Glycidyl methacrylate (GMA) was chosen as a monomer for the synthesis of the brush backbone; this monomer meets the requirements for the both *click* chemistry reactions. Side chains were synthesized by polymerization of *tert*-butyl methacrylate (*t*-BMA). After modification, GMA-based backbone and p(*t*-BMA) side chains were coupled into polymer brushes via epoxy-thiol or azide-alkyne *click* reactions. The course of the *click* reactions was monitored by DLS. After acidic hydrolysis of p(*t*-BMA) chains with TFA, anionic polymer brushes were obtained.

Solution properties of the anionic polymer brushes were investigated by potentiometric/turbidimetric titrations and DLS. It was determined that the both molecular brushes, irrespective of the type of *click* chemistry reaction, were weaker polyacids compared to linear poly(methacrylic acid) (pMAA, pK_{α} 6.5-6.7). At high pH values (10-12), anionic polymer brushes are deprotonated, fully stretched and are individually solubilized (R_h ~12-19 nm) (Fig. 1). In close to neutral solutions (pH 6–8), the side chains of pMMA become partly protonated, less soluble, and tend to compact. Full protonation is reached at about pH ~5, where medium-size micelles are formed (R_h 45-60 nm). In acidic solutions (pH <4), large micellar aggregates are formed which are unstable and susceptible to sedimentation.

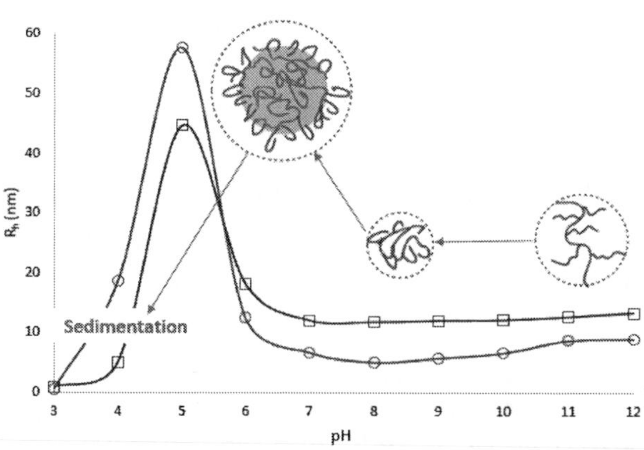

Fig. 1. Hydrodynamic radius of anionic polymer brushes *vs.* solution pH: —○— synthesis by azide-alkyne *click* chemistry, —□— synthesis by epoxy-thiol *click* chemistry.

References

1. H. Lee et al., Prog. In Polym. Sci., **35**, 24-44 (2010).
2. G. Kocak et al., Polym. Chem., **8**, 144-176 (2017).
3. Y. Gao et al., Macromol. Res., **6**, 513-527 (2017).

INVESTIGATION OF CONJUGATED WATER SOLUBLE POLYMER MPS-PPV

M. Jurkūnas[1,2], **A. Stirkė**[1], **A. Vareikis**[2]

[1] *Department of Material Science and Electrical Engineering, Center for Physical Sciences and Technology, Lithuania, marijus.jurkunas@chf.stud.vu.lt*

[2] *Department of Polymer Chemistry of Faculty of Chemistry and Geosciences, Vilnius University, Lithuania*

Conjugated polyelectrolytes (CPEs), such as poly[5-methoxy-2-(3-sulfopropoxy)-1,4-phenylenevinylene] (MPS-PPV), are organic semiconductors and polyelectrolytes at the same time. Recently they have attracted increasing attention for their potential applications in optoelectronic devices, in bioimaging [1, 2], in chemo- and biosensing [3], or have been extensively studied as photo-active layers in organic light emitting diodes, organic solar or organic photovoltaic cells [2]. In spite of the fact that MPS-PPV is well known, versatile and widely studied polymer, some basic properties of it have been poorly elucidated earlier. Therefore this study is intended somewhat to fulfil this gap as well as to provide some new insights on the possible in future applications of MPS-PPV.

Based on the Gilch dehydrohalogenation reaction [3], the MPS-PPV has been synthesized and fully characterized. The structure of polymer and its precursors were confirmed by ^{1}H NMR and FT-IR spectroscopy. TGA analysis showed that polymer contains about 7-8% of humidity and sustains 2-step thermal degradation. Knowing that MPS-PPV is a negatively charged polyelectrolyte and fluorescent in the UV light, a new method for its molecular weight determination was developed by monitoring electrophoretic mobility in agarose gel of both MPS-PPV and Gene Ruler, 1 kb plus DNA ladder standard. Accordingly, 20-60 kDa molecular weights seem to be most dominant but fractions up to 240 kDa are also observable. On the other hand, centrifugal filtration test with 5 different molecular weight cutoffs showed that polymer mostly comprises of 100-300 kDa fractions meaning that MPS-PPV mobility in agarose is higher than that of DNA standard. Fluorescence of MPS-PPV in DMSO is 20 times stronger than in aqueous solution. Most intense fluorescence obtained at concentration ~10^{-4} M in both solutions. Reabsorption effect of MPS-PPV can be observed only in DMSO solution when diluting it from 10^{-3} to 10^{-4} M. Fluorescence quenching experiment showed 3-step decay mechanism for water solution and 2-step for DMSO solution. Upon dilution, quantum efficiency of MPS-PPV solutions grows form ~7% to ~16%. Experimentally determined HOMO and LUMO energy levels of MPS-PPV were ~5.2 eV and ~3.0 eV, respectively, indicating that MPS-PPV is perfectly suitable for building of polymer light-emitting diode (PLED).

We have built FTO/TPD/MPS-PPV/Alq3/Al PLED which showed a weak red-orange glow, while slightly modified structure ITO/TPD/MPS-PPV/Alq3/LiF/Al showed only Alq3 electroluminescence spectra. We also tested MPS-PPV as a dopant for chemically prepared polypyrrole-poly(vinyl alcohol) (PPy-PVA) matrix with the idea of making such matrix more conductive and possibly electrically stable. Electrochemically deposited PPy layer could be also doped with the MPS-PPV using simple solution of both compounds and cyclic voltammetry technique (0 V to +0.8 V, 100 cycles at 50 mV/s). Such layer appeared 200 times thinner, smoother and far stronger attached to ITO electrode than PPy layer obtained in the absence of MPS-PPV. This experiment has shown tremendous potential to solve adhesiveness problems of electrochemically deposited conjugated polymers on ITO.

References

1. Z. Kahveci et al., ACS Applied Materials & Interfaces, **8**, 1958□1969 (2016).
2. L. Chen et al., Proceedings of the National Academy of Sciences, **96**, 12287□12292 (1999).
3. J. Dalvi-Malhotra et al., The Journal of Physical Chemistry B, **109**, 3873□3878 (2005).

SYNTHESIS AND STUDY OF CATECHOL GROUPS CONTAINING COPOLYMERS

M. Steponavičiūtė, V. Klimkevičius, R. Makuška

Institute of Chemistry, Vilnius University, Naugarduko str. 24, 03225 Vilnius, Lithuania, *steponaviciute.medeina@gmail.com*

Mussels are famous for the ability to survive in harsh marine conditions by their robust attachment to various substrates. The "sticky" attachment apparatus is known as byssus, which is formed from the secretion of liquid mussel foot proteins. It is known [1] that excellent adherence properties of blue mussels in wet conditions are predetermined by catechol-containing amino acid, 3,4-dihydroxyphenylalanine. Many catechol groups containing polymers were synthesized as adhesives, anticorrosion polymer coatings, and surface modifiers [2, 3]. Between these polymeric materials, there are only very few examples of the brush copolymers carrying catechol groups [4].

First, in the synthesis of catechol-functionalized polymers, special care needs to be taken due to their chemical reactivity. Catechol group containing monomers act as weak inhibitors of radical reactions, and are readily oxidized to quinones. In order to avoid undesirable reactions, two approaches were used – protection of the catechol group present in a monomer or introduction of moieties with catechol group by modification of random copolymers.

Random brush copolymers were synthesized by RAFT copolymerization of poly(ethylene oxide) methyl ether methacrylate (PEOMEMA, M_n 950) with acetonide protected dopamine methacrylamide (A-DOPMA) (the first approach), or PEOMEMA with DMAEMA (the second approach). The protecting groups of A-DOPMA units were removed using trifluoroacetic acid. DMAEMA units in the copolymers were quaternized with 2-chloro-3,4-dihydroxyacetophenone to give QDMAEMA. The copolymers of PEOMEMA and DOPMA are catechol groups containing brush copolymers without charge, and the copolymers of PEOMEMA and QDMAEMA are catechol groups containing brush copolymers carrying positive charge.

Copolymerization kinetics and the copolymers were studied by size exclusion chromatography with triple detection (RI, LS, and DP) and ^{1}H-NMR spectroscopy. Degree of polymerization of the copolymers was close to 100, and the dispersity index Đ was less than 1.2. Resistance to oxidation of the copolymers with catechol groups was studied by UV-Vis spectroscopy, and nanoscale wear resistance of the adsorbed layers by AFM-based methodology. The copolymers with protected catechol groups were rather stable, while those with deprotected catechol groups were sensitive to oxidation and UV irradiation. The most unstable were cationic catechol groups containing brush copolymers.

Acknowledgements. In collaboration with Prof. Per M. Claesson, KTH, Sweden.

References

1. M. Yu, J. Hwang, and T. Deming, J. Am. Chem. Soc., **121**, 5825–5826 (1999).
2. J. Yang et al., Macromol. Chem. Phys., **219**, 16, 1800051 (2018).
3. H. Lee, B.P. Lee, and P.B. Messersmith, Nature, **448**, 338–341 (2007).
4. N. Patil et al., Polym. Chem., **6**, 2919–2933 (2015).

SYNTHESIS AND STUDY OF PENTABLOCK COPOLYMERS BY CONSERVATIVE AND ONE-POT CHAIN EXTENSION RAFT POLYMERIZATION

E. Kubricenko, T. Krivorotova

Institute of Chemistry, Vilnius University, Vilnius, Lithuania, tania.krivorot@gmail.com

It was shown recently [1] that RAFT polymerization is well suited for the synthesis of high-order multiblock copolymers by one-pot sequential monomer addition, where each step is taken to near full monomer conversion without intermediate purification. In the present study, one-pot methodology was used for the synthesis of a series of hydrophilic pentablock copolymers with rather long blocks (up to DP 80) differing by the nature of the monomers as well as succession of their blocks.

Amphiphilic pentablock copolymers containing the blocks of dimethylamino ethyl methacrylate (DMAEMA) and (polypropylenglycol) methacrylate (PPGMA) were synthesized by conservative and one-pot RAFT polymerization using difunctional RAFT CTA, which allowed the extension of growing chains into two directions and the synthesis of pentablock copolymers in three steps [2]. The RAFT polymerization was studied using two chain transfer agents: ethylene glycol di((1-butyl)sulfanylthiocarbonylsulfanyl-4-cyanopentanoate) or its difunctional analog with dithioben-zoate groups in the mixture of dioxane and *N*-metyl-2-pyrrolidone or in the mixture of butanol. Conversion of the monomers was monitored by ^{1}H NMR spectroscopy, and molecular weight of the copolymers was determined by SEC with triple detection. Trying to obtain "pure" blocks, polymerization was carried out up to very high conversions (usually, over 98 mol.%), and then the next monomer was added.

Pentablock copolymers containing the blocks of pDMAEMA and pPPGMA (at $T < 10^oC$) are well soluble in water, while containing the same blocks of pDMAEMA and pPPGMA at $T > 10^oC$ could be solubilized in organic solvents only [3]. The copolymers containing pPPGMA blocks with rather long PPG side chains represent molecular brushes while those containing pDMAEMA blocks are cationic in acidic media.

The LCST transition of copolymers containing either HEMA-DMAEMA or PEOMEMA-DMAEMA monomers couples was investigated in detal by DLS measurements as function of temperature and pH for two pentabloks containing at the beginning of copolymer either DMAEMA or PPGMA. The temperature induced the agglomerates of copolymer PPGMA-DMAEMA-PPGMA-DMAEMA-PPGMA at lower temperatures that in aqua solution of DMAEMA-PPGMA-DMAEMA-PPGMA-DMAEMA copolymer.

References

1. G. Gody, T. Maschmeyer, P. B. Zetterlund, S. Perrier, Macromolecules., **47**, 3451–3460 (2014).
2. T. Krivorotova, P. Radzevicius, and R. Makuska, Eur. Polym. J. **66**, 543-557 (2015).
3. A. A. Steinschulte,W. Xu, F. Draber, P.Hebbeker *et al.,* Soft Matter, **11**, 3559-3564 (2012).
4. A. A. Steinschulte, B. Schulte, S. Ruetten, T. Eckert, Phys. Chem. Chem.Phys., 16, 4917 (2014).

MAXWELL EFFECT IN SOLUTIONS OF AMPHIPHILIC COPOLYMERS BASED ON N-METYL-N-VINYLACETAMIDE

A. Gosteva, G. Kolbina

Institute of Macromolecular Compounds, Russian Academy of Sciences, St. Petersburg, Russia, anna.gosteva.imc@gmail.com

Nowadays, intensive efforts are being made to synthesize and investigate polymer-carriers, that is, molecular containers for targeted transport of various biologically active substances. The amphiphilic copolymers are suitable for these purposes, i.e. they contribute the formation of unimolecular micelles. One of the main requirements for such systems is biocompatibility, which is characterized by the absence of toxicity and effective functioning during use. However, for the majority of synthesized container systems, the information on their structure and structure–property relationships is very limited or even absent. Thus, the synthesize and investigate of polymer-carriers for the delivery of biologically active substances based on amphiphilic polymers is a relevant and promising direction in biotechnology. The alkylated copolymers based on N-methyl-N-vinyl acetamide, which are one of candidates to such macromolecular systems.

Such macromolecules have hydrophilic backbone and hydrophobic side radicals. Therefore MVAA-MVAH alkylated copolymers might be considered as potential drug delivery systems. Samples were investigated by molecular optics - flow birefringence (Maxwell effect) in 0.1 M NaCl with suppression of polyelectrolyte effects. This method is very sensitive to the structure and conformation of the macromolecule.

Noticeable differences in the properties of copolymers with short and long alkyl groups have been established.

Acknowledgements. The reported study was funded by RFBR according to the research project № 18-33-00432.

SYNTHESIS OF N-VINYLSUCCINIMIDE COPOLYMERS WITH VINYL ACETATE AND N-VINYLPYRROLIDONE UNDER REVERSIBLE CHAIN TRANSFER CONDITIONS

A. I. Gostev, D. V. Grigoriev, S. A. Satarova, E. V. Sivtsov

Saint-Petersburg State Institute of Technology, Saint-Petersburg, Russia, dmitriy-get@rambler.ru

The copolymers of N-vinylsuccinimide (VSI) have attracted attention for several decades as the basis for preparation of polymer matrices of medicobiological use. By means of alkaline or acid hydrolysis VSI units can be easily transformed into the units of N-vinylamidosuccinic acid (VASA), to which low-molecular-weight compounds of the base nature (e.g. amines) can be attached via ionic bond. Traditional radical copolymerization of VSI is well studied. A new wave of interest to these objects is associated with the development of techniques of controlled radical polymerization, primarily polymerization under the conditions of reversible chain transfer – RAFT polymerization.

Copolymers VSI with vinyl acetate (VA) and N-vinylpyrrolidone (VP) are already known; the values of reactivity ratios under various conditions are shown in the Table. The use of RAFT polymerization can significantly improve the molecular mass characteristics of these polymers, which is important for their further application as part of macromolecular medicine forms, and also makes it possible to obtain compositionally homogeneous copolymers, despite a deep difference in the relative activities of monomers.

Table. Monomers reactivity ratios in copolymerization of N-vinylsuccinimide (M_1) with vinyl acetate and N-vinylpyrrolidone.

Comonomers	Solvent	Initiator	T, °C	r_1	r_2
vinyl acetate	In bulk	BP	65	5.1	0.18
	Ethanol	BP	70	5.62	0.17
	DMSO (VSI:DMSO=1:3 mol.)	AIBN	70	2.78	0.04
	DMSO (VSI:DMSO=1:14 mol.)	AIBN	70	0.02	0.82
N-vinylpyrrolidone	In bulk	AIBN	45	1.30	0.41
	Ethanol	AIBN	60	1.50	0.40

In this study, we solve the problem of copolymer synthesis in the presence of dibenzyltrithiocarbonate (BTC) as a RAFT agent, which is effective only with respect to VSI ($C_{tr} \sim 19$). Homopolymerization of VA in the presence of BTC does not occur at all, and VP – with insufficient efficiency of control of MM characteristics. Nevertheless, the achievement of the goal is possible due to the fact that VSI is more active in the both monomer pairs studied (see the Table), and therefore the *effective* value of the chain transfer constant for copolymerization must be higher than in homopolymerization of VA and VP. In addition, it is known that polymeric RAFT agents are more effective than the initial low molecular weight one. Therefore, it is possible to use copolymers obtained at high concentrations of BTC as polymeric RAFT agents. This allowed us to synthesize copolymers with high yield and good control of molecular mass characteristics. The obtained copolymers VSI–VA and VSI–VP were characterized by NMR, IR spectroscopy, and gel permeation chromatography. It has been shown that polymers with narrow unimodal MWD have been obtained. On the basis of these precursors, alkaline hydrolysis of the corresponding copolymers of VASA–vinyl alcohol and VASA–VP have been obtained and their complexes with rimantadine have been prepared.

Acknowledgements. This work was supported by the Ministry of Science and Education of the Russian Federation (11.5362.2017/8.9).

CONTROLLED CATIONIC POLYMERIZATION OF *TRANS*-ANETHOLE

M. I. Hulnik[1,2], **O. V. Kuharenko**[1,2], **I. V. Vasilenko**[1], **S. V. Kostjuk**[1,2]

[1] *Research Institute for Physical Chemical Problems of the Belarusian State University, Minsk, Belarus, maksim.gulnik@gmail.com*
[2] *Department of Chemistry, Belarusian State University, Minsk, Belarus*

Owing to the expected exhaustion of fossil oil reserves and environmental concerns, synthetic polymers from bio-derived compounds seem to be a promising alternative materials to replace widely used polymeric materials obtained from petrochemicals. Herein, we report the controlled cationic polymerization of *trans*-anethole – a derivative of phenylpropene extensively occurring in essential oils of different plants, such as anise and fennel [1]. The cationic polymerization of *trans*-anethole was investigated using three-components initiating systems CumCl/2,6-lutidine/$SnCl_4$ and p-MetOStCl/2,6-lutidine/$SnCl_4$ in toluene at -60 °C (Figure 1).

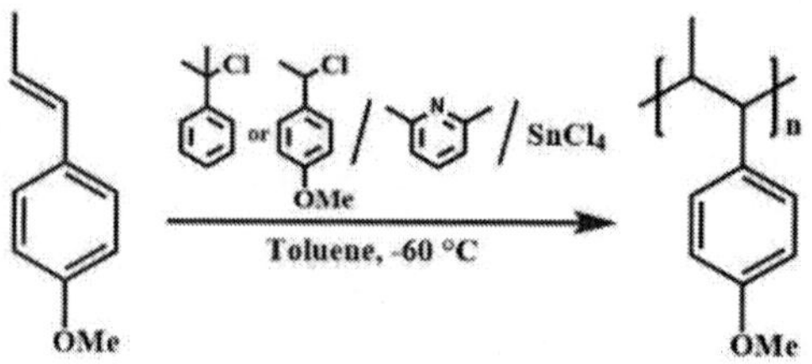

Fig. 1. Cationic polymerization of *trans*-anethole

The $SnCl_4$-coinitiated cationic polymerization of *trans*-anethole proceeds in a controlled fashion that was confirmed by the linearity of first order kinetic plots, linear growth of molecular weight with monomer conversion and relatively narrow polydispersity of the obtained polymers ($M_w/M_n = 1.26\text{-}1.45$). In order to confirm the controlled character of cationic polymerization of *trans*-anethole, the different monomer to initiator ratios were examined. In the present work, poly(*trans*-anethole)s with controlled molecular weight from 5000 to 15000 were synthesized and characterized using GPC, ^1H NMR and MALDI-TOF methods. SEC traces of the synthesized poly(*trans*-anethole)s along with the main peak showed a small fraction (<10%) in high molecular weight region. Despite the formation of high molecular weight fraction, SEC traces shifted to high molecular weight region confirming controlled character of the polymerization process. It was shown by MALDI-TOF analysis that poly(*trans*-anethole)s synthesized using CumCl as initiator contain two types of head groups, while only one type of head group, which corresponds to structure of initiator, is formed in the case of using p-MetOStCl as an initiator.

Acknowledgements. This work was supported by the State Belarusian Program of Scientific Research "Chemical technologies and materials" subprogram «Wood chemistry», project 4.1.12.

References

1. B.B. Bharat et al., Biochem. Pharmacol., **71(10)**, 1397–1421 (2006).

ELECTROCHEMICAL PERFORMANCE OF HYBRID POLYPYRROLE / Fe_2O_3 HYDROGEL

V. K. Vorobiov[1], M. P. Sokolova[1], A. N. Bugrov[1,3], S. N. Bolshakov[3], I. A. Kasatkin[2], M. A. Smirnov[1]

[1] *Institute of Macromolecular Compounds, Russian Academy of Sciences, Saint Petersburg, Russia, vrbvrbvrb@mail.ru*
[2] *Saint Petersburg State University, Saint Petersburg, Russia*
[3] *Saint Petersburg Electrotechnical University "LETI", Saint Petersburg, Russia*

Supercapacitors and pseudo-supercapacitors based on electroconductive polymers (EPs), such as polypyrrole and polyaniline, and containing redox-active nanoparticles (RuO_2, Fe_2O_3, MnO_2) attracted a great interest during the last decades due to a high specific capacitance and a stable cycle life performance [1]. In such organic-inorganic composites EPs enhance electrical contact area between oxides nanoparticles and current collector. At the same time, the EP-based electroconducting hydrogels used as a matrix for the electrodes allow increasing the active mass loaded on the electrode due to the increased ion mobility [2]. Thus, the hybrid (organic-inorganic) hydrogels can combine the benefits of the high ion mobility and electrical conductivity of EPs with the high theoretical pseudocapacitance of inorganic oxides.

In this work, the hybrid hydrogel was prepared by oxidative polymerization of pyrrole in the presence of Fe_2O_3 nanoparticles using sodium trimetaphosphate as cross-linker. A sample without adding of nanoparticles was prepared under the same conditions for comparison. Wide-angle X-ray diffraction characterization of the samples was performed in both dry and swollen states. The chemical structure of the hydrogels was studied with FTIR spectroscopy. Electrochemical properties of the hydrogels were investigated by cyclic voltammetry, galvanostatic charge-discharge, and impedance spectroscopy measurements in a 3-electrode cell with 1M Na_2SO_4 solution as electrolyte. The results show higher specific capacitance of the PPy-Fe_2O_3 hydrogel - up to 509 F/g at the current density of 0.2 A/g - in comparison with 250 F/g for the pure PPy hydrogel. The cycling stability after applying 3000 charge-discharge cycles was similar for both samples that indicates a good compatibility of the components in the composite.

Acknowledgements. This work was supported by Russian Foundation of Basic Research (grant 18-03-01167 A). The experimental work was facilitated by the equipment of the Resource Centre of X-ray Diffraction Studies at St. Petersburg State University.

References

1. I. Shown et al., Energy Science & Engineering., **3**, № 1, 2– 26, (2015).
2. M.A. Smirnov et al., Materials Letters., **199**, 192–195 (2017).

GRAFTED PENTABLOCK-COPOLYMERS WITH MIXED LINEAR-BRUSH TOPOLOGY PMMA-BLOCK-PCL-BLOCK-(PI-GRAFT-PMMA)-BLOCK-PCL-BLOCK-PMMA

A. Kashina[1], T. Meleshko[1], N. Bogorad[1], M.Bezrukova[1], A.Yakimansky[1,2]

[1] *Institute of Macromolecular Compounds of Russian Academy of Sciences, St.Petersburg, Russia, kashina.anna@mail.ru*
[2] *Institute of Chemistry, Saint Petersburg State University, St.Petersburg, Russia*

One of the main trends in the modern polymer chemistry is to synthesize polymers with more and more complex architecture, which allows obtaining materials with previously unattainable properties. For example, triple multiblock-copolymers with mixed linear-brush topology with a central brush-type block and outer linear blocks is prominent for creating synthetic supersoft elastomers. According to this trend, our work is devoted to development of methods for the synthesis of multi-block copolymers with different chemical structure of individual blocks.

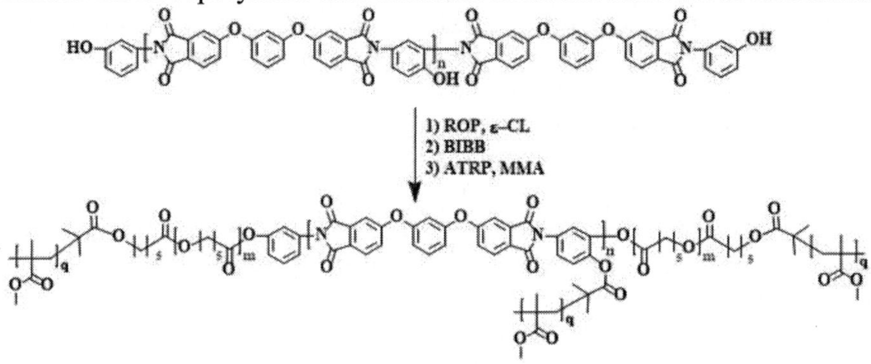

Fig. 1. The scheme of the synthesis of grafted pentablock-copolymers PMMA-block-PCL-block-(PI-graft-PMMA)-block-PCL-block-PMMA.

The preparation of grafted copolyimides with mixed linear-brush topology was carried out in several stages. First, multicenter polyimide initiators were synthesized with phenol groups both in each repeating unit and at the ends of polymer chains. Second, ε-caprolactone (CL) was polymerized by the ROP on this macroinitiators. As a result, linear block copolymers of the ABA type were obtained, where block B was polyimide with phenol groups in each repeating unit, and block A was PCL. In order to achieve brush-type structure of central block, the obtained linear triblock copolymers were functionalized by esterification of hydroxyl groups located in each PI unit and at the ends of the PCL chains. Then methyl methacrylate (MMA) was polymerized using the ATRP method on the functionalized multi-center triblock-copolymer initiators. The chemical structure of the obtained copolymers was proved by ^{1}H NMR and IR spectroscopy. The molecular weights (MW) of the synthesized copolymers were determined by SEC with triple detection. The values of MW of triblock- and pentablock-copolymer with a brush-type central block were additionally determined by sedimentation-diffusion analysis. The obtained absolute MWs were in good agreement with the SEC data. Thus, it was shown that the combination of ATRP and ROP methods using multifunctional polyimide macroinitiators allows to synthesize copolymers with complex topology with central brush-type block of graft copolyimide with PMMA side chains terminated with PCL–block–PMMA linear block copolymers.

Acknowledgements. The work was supported by the Russian Ministry of Education and Science [grant number 14.W03.31.0022 (Megagrant of the Government of the Russian Federation)].

DESIGN, SYNTHESIS AND SELF-ASSEMBLY OF AMPHIPHILIC MULTICOMPONENT MOLECULAR POLYIMIDE BRUSHES

I. Ivanov[1], M. Simonova[1], A. Kashina[1], T. Meleshko[1], A. Filippov[1], A. Yakimansky[1,2]

[1] *Institute of macromolecular compounds of the Russian academy of sciences, St. Petersburg, Russia, gangspil@gmail.com*
[2] *Saint Petersburg State University, Institute of Chemistry, St. Petersburg, Petergof, Russia*

The synthesis and study of amphiphilic polymer systems of complex architecture promising for the engineering and creation of new materials has been actively developed in recent years. Achieving the precise control of the structure, topology and distribution of blocks of different nature in macromolecules of complex architecture is a complicated synthetic task, but provides extensive opportunities for designing new functional materials with specified characteristics. Molecular brushes (MBs) (densely grafted copolymers) are the subject of particular interest due to their unusual architecture and a set of unique chemical and physical properties.
The scope of the present research is development of methods for the synthesis of amphiphilic MBs with a hydrophobic polyimide (PI) backbone (polyimide brushes) and amphiphilic diblock-copolymer side chains of polymethacrylic acid and poly(methyl methacrylate) (PMAA-b-PMMA).

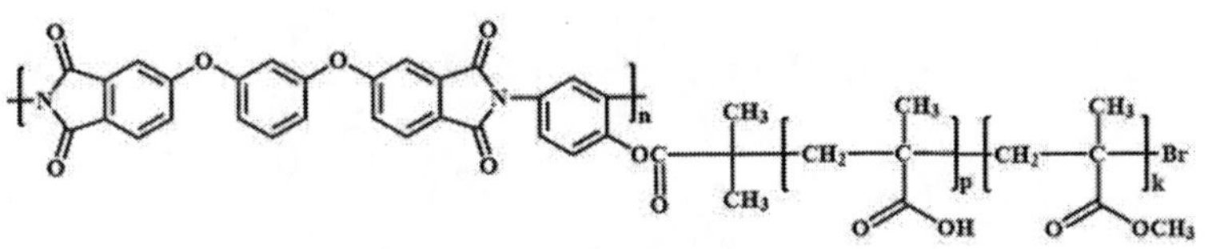

Synthesis of target MBs with block-copolymer side chains was carried out by the "grafting from" approach in conjunction with ATRP method in several stages through the intermediate formation of MBs with regularly grafted side chains of poly(*tert*-butyl methacrylate) (PTBMA) followed by chain extension of MMA from living chain ends of PI-g-PTBMA. At the last stage amphiphilic MBs with hydrophilic PMAA blocks in side chains were obtained by acidic hydrolysis of TBMA units. Based on the analysis of the polymerization kinetics and molecular weight characteristics of polymerization products the conditions allowing regulation of grafting density, length and distribution of hydrophilic and hydrophobic blocks in side chains were determined.
The self-assembly of synthesized amphiphilic MBs in selective solvents was investigated by DLS and TEM. It was shown that PI-*g*-(PMAA-*b*-PMMA) brushes tend to self-assembly in ethanol with formation of stable supramolecular nanostructures. The influence of structural parameters of MBs (grafting density, length of side chains and ratio of hydrophilic and hydrophobic segments of macromolecules) on conformational characteristics of self-assembled nanostructures was established. The possible mechanism of self-assembly of MBs in selective solvent was proposed.

Acknowledgments. The work was supported by the Russian Ministry of Education and Science [grant number 14.W03.31.0022 (Megagrant of the Government of the Russian Federation)].

BEHAVIOR OF N-METHYL-N-VINYLACETAMIDE AND N-METHYL-N-VINYLAMINE HYDROCHLORIDE ALKYLATED COPOLYMERS IN DIFFERENT SOLVENTS

O. Dommes, **A. Gosteva, O. Okatova, G. M. Pavlov**

Institute of Macromolecular Compounds, Russian Academy of Sciences, St. Petersburg, Russia, olga.dommes.imc@gmail.com

Polymers based on *N*-vinylamides are actively used in many areas of technology and medicine, for example, as flocculants for water treatment and dehydration of residues in paper production, in cosmetics etc. In this class of polymers special role play N-vinilacetamide based polymers due to large complex of pharmaceutical activities: biological, immunomodulating and antithrombogenic ones [1]. By partial hydrolysis of poly-*N*-vinylacetamides, it is possible to control the charge of a macromolecule and by additional alkylation – to control hydrophobic-hydrophilic balance and thus obtaining molecular containers for targeted transport of various medicines.

N-methyl-N-vinylacetamide and N-methyl-N-vinylamine hydrochloride (MVAA-MVAH) alkylated copolymers were studied by the methods of molecular hydrodynamics: velocity sedimentation, translational diffusion, and viscometry. The copolymer structure is presented in Fig. 1.

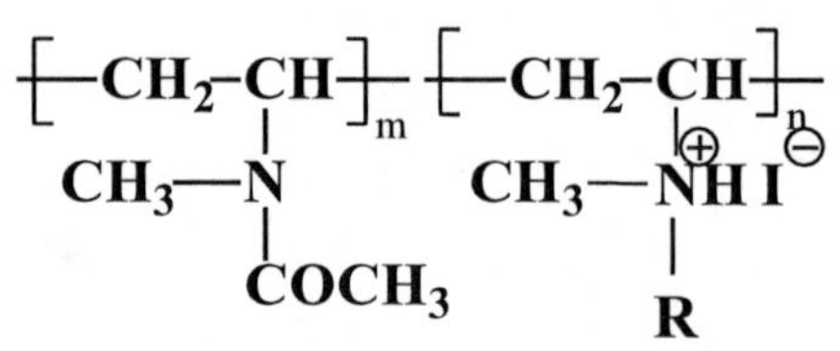

R = $-C_6H_{13}$; $-C_8H_{17}$; $-C_{10}H_{21}$; $-C_{12}H_{25}$

Fig. 1. The structure of N-methyl-N-vinylacetamide and N-methyl-N-vinylamine hydrochloride alkylated copolymers

Such macromolecules have hydrophilic backbone and hydrophobic C_6H_{12}–$C_{12}H_{25}$ side radicals and possibly may form hydrophobic internal domains in macromolecular coils. Therefore MVAA-MVAH alkylated copolymers might be considered as potential drug delivery systems. Due to presence of charged groups competition of electrostatic and hydrophilic-hydrophobic interactions takes place. The prevalence of some type of such interactions depends on the ionic strength of solution. Previous viscometry titration measurements demonstrated [2] the compaction and possible precipitation of macromolecule at high ionic strengths (0.5–1.6 M NaCl). In this work sedimentation–diffusion and viscometry analysis in a wide range of ionic strength were carried out. The conformational properties of MVAA-MVAH alkylated copolymers in different solvents are discussed.

Acknowledgements. The reported study was funded by RFBR according to the research project № 18-33-00432.

References

1. E. F. Panarin et al., Polymers – Carriers of Biologically Active Substances, Professija, St. Petersburg (2014).
2. O. A. Dommes et al., IOP Conf. Ser.: Mater. Sci. Eng., **500,** 012017 (2019).

COPOLYMERS N-SUBSTITUTED 2-AZANORBORNENES WITH ACRYLIC ACID: SYNTHESIS AND BIOMEDICAL APPLICATION

M. N. Gorbunova

Institute of Technical Chemistry, Ural Branch of Russian Academy of Sciences, Korolev str., 3, Perm 614013, Russia mngorb@yandex.ru

Novel copolymers of N-methyl-2-azanorbornene-5 (MAN), N-benzyl-2-azanorbornene-5 (BAN), N-allyl-2-azanorbornene-5 (AAN) and N-(2-azanorbornene-5-en)methylacetate (MAAN) with acrylic acid (AA) were synthesized in bulk by the method of radical copolymerization in the presence of radical initiator AIBN.

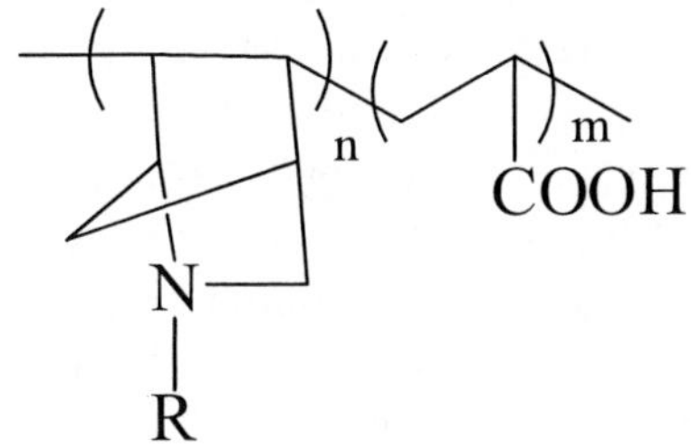

R = CH_3 (МАН), $C_6H_5CH_2$ (БАН), CH_2=$CHCH_2$ (ААН), CH_3OCOCH_2 (МААН)

It is established that the copolymerization of azanorbornenes (ANB) proceeds with the formation of statistical copolymers with high tendency of monomer units to alternation. The kinetic regularities of the reactions were investigated and it was found that the copolymerization reaction rates decrease with an increase of azanorbornene content in the initial monomer mixture.

By ^{13}C NMR, it was established that the copolymerization of azanorbornenes with N-vinylpyrrolidone proceeds through the double bond of the norbornene ring. The allyl group of N-allyl-2-azanorbornene-5 does not participate in the copolymerization both at the equimolar ratio of monomers and at the double excess of acrylic acid.

Copolymers ANB with AA obtained are dark powders and are soluble in methanol, DMSO, dimethylformamide. Copolymers ANB with AA were found to be nontoxic (the LD_{50} values were more 1000 $mg \cdot kg^{-1}$) and therefore could be used for medical purposes. The study of the cytotoxic activity of the copolymers showed that the copolymer of AAN with AA has activity against MS melanoma cells; copolymer at a concentration of 32.49 μg/ml inhibits 50% of melanoma cells.

Acknowledgement: Financial support by the Russian Foundation for Basic Reseach (grant № 19-43-590019-r_a) is gratefully acknowledged.

SYNTHESIS OF HIGHLY REACTIVE POLYISOBUTYLENE FROM C_4 MIXED FEED USING CHLOROFERRATE IMIDAZOLE-BASED IONIC LIQUID AS CATALYST

I. A Berezianko[1,2], **I. V. Vasilenko**[2], **D. I. Shiman**[2], **S. V. Kostjuk**[1,2,3]

[1] *Belarusian State University, Department of Chemistry, Minsk, Belarus,* bereziankoivan@bsu.by
[2] *Research Institute for Physical Chemical Problems of the Belarusian State University, Minsk, Belarus*
[3] *Sechenov First Moscow State Medical University, Institute for Regenerative Medicine, Moscow, Russia*

Highly reactive polyisobutylene (HR PIB) represents an important class of synthetic industrial polymers, which found an application as intermediate in the manufacturing of motor oils and fuel additives [1]. Such type of polymers is characterized by low molecular weight ($M_n \sim 500$–5000 g/mol) and possessed mainly *exo*-olefin end groups at a chain end. Nowadays HR PIB is commercially produced from isobutylene (IB) or C_4 mixture feed, which contains IB (≈46 wt%), 1-butene (≈24 wt%) and 2-butenes (≈17 wt% *cis* and *trans* isomers) and also lower alkanes as a residue. Against the background of numerous developed methods of HR PIB producing a special place occupies a cationic polymerization of IB co-initiated by acidic imidazole based ionic liquids [2,3].

In this work we expand the borders of application of ionic liquids (ILs) as catalysts for the preparation of HR PIB from C_4 mixed feed. Here we used IL based on 1-ethyl-3-methylimidazolium chloride and $FeCl_3$ in combination with diisopropyl ether. The obtained results applying of this catalytic system (solvent – *n*-hexane) are summarized in Table 1.

Firstly, the surprisingly high *exo*-olefin end group content in the absence of $^iPr_2O$ in the polymerization system should be noted. On the other hand, the decrease of polymerization rate in comparison to pure IB in similar reaction condition is observed [3]. Therefore, for reaching high monomer conversions higher IL concentration (44 mM) or prolonged reaction time (60 min) are required (Table 1). Taking into account heterogeneous character of IL co-initiated polymerization in *n*-hexane, sonication of reaction mixture during 3 minutes before C_4 mixture feed addition was applied in attempt to increase the reaction rate. As a result, polymerization rate increased in 2 times similarly to polymerization of pure IB [3]. It should be also noted that relatively narrow molecular-weight distribution ($M_w/M_n \leq 2.8$) and high *exo*-olefin end group content (≥75%) were observed (Table 1). Thus, ILs is quite promising catalyst for the preparation of HR PIB via cationic polymerization of C_4 mixed feed.

Table 1. Polymerization of C_4 mixed feed at 0°C catalyzed by ionic liquid ([IB] = 3.8 M).

№	[IL], mM	[$^iPr_2O$], mM	Sonication	t, min	IB conv, %	M_n (SEC), g/mol	M_w/M_n	exo-olefin end groups, %
1	22	-	NO	30	18	5800	3.5	74
2	22	11	NO	30	19	1700	2.7	83
3	33	11	NO	30	24	2200	2.8	88
4	33	11	YES	30	51	2200	2.8	84
5	33	11	YES	60	78	2500	2.5	81
6	44	11	NO	30	68	3900	2.6	82
7	44	11	NO	60	88	3700	2.6	75

Acknowledgements. The authors thank BASF SE for the continuous financial support of this research.

References

1. H. Mach and P. Rath, Lubr. Sci., **11-2**, 175–185 (1999).
2. I. Vasilenko et al., Polym. Chem., **7**, 5615–5619 (2016).
3. I. Berezianko et al., Polymer, **145**, 382–390 (2018).

COATING FOR REDUCING THE FLAMMABILITY OF LINEN FABRIC

A. Bernava

Riga Technical University, Institute of Polymer Materials, Riga, Latvia, aina.bernava@rtu.lv

Most of the fibre-forming polymers have a common problem, i.e. majority of them burn under normal environmental conditions and therefore present a serious fire hazard. Conventionally, to textiles generally are applied flame-retardants, either by impregnating them in a solution containing a flame-retardant as well or by coating or spraying flame-retardant to one side of the fabric (back coating) [1].

For obtaining durable flame-proof and flame-retardant properties of the fire-retardant materials in production process a lot of chemicals, such as ammonium polyphosphates, urea, polyphosphoric acid, organic-phosphate resins and their mixes [2] can be used. The antimony and antimony compounds are safely produced and used in the textile industry without causing a risk to human health and environment [3]. Examples of flame-retardant coated fabrics application is furniture upholstery, roller-blinds, interior materials for transportation, protection garments etc. [4].

In our research the flat screen coating method is used for treatment of raw and bleached linen fabric (Larelini Ltd., Latvia). The commercial printing pastes Tubicsreen EX-TS and Printperfect 226 (producer CHT BEZEMA) and 4% additive antimony trioxide (Sigma-Aldrich); drying of coatings at 100°C for 7 min and thermal treating at 160°C for 5 min. are applied. The right side of fabric is coated with continuous coating which is preferable.

For flammability tests LVS EN ISO 15025:2003 horizontal method is adapted. Coated textiles are examined from their right and left sides. The abrasion resistance of untreated and coated samples on Taber Rotary Platform Abraser, according to LVS EN ISO5470-1:2001 is determined.

The results of the test show, that the raw and the bleached linen fabrics without coating burn down completely. The flame retardancy of all coated fabrics improves as well as increase of abrasion resistance is. The best results are for coatings with commercial printing paste Tubicsreen EX-TS, containing the antimony trioxide.

References

1.M. Neisiuset al., Functional Finishes for Textiles, Woodhead Publishing Series in Textiles 429-461 (2014).

2.S. M.Mostashariet al., J. of Therm.An. and Calor. **95,1**, 187-192 (2009).

3. Flame Retardants Workshop, EFRA March 3-4 (2015).

4.W. Fung, Coated and Laminated Textiles. Woodhead Publishing, 72 (2002).

BIODEGRADABLE MULTILIPOSOMAL CONTAINERS BASED ON CHITOSAN

E. Krasnikov[1], A. Efimova[1], G.Krivtsov[2], G. Rudenskaya[1], A. Yaroslavov[1]

[1] *Polymer Department, Faculty of Chemistry, M.V.Lomonosov Moscow State University, Moscow, Russia, soloveyteam@mail.ru*
[2] *Mechnikov Research Institute of Vaccines and Sera, Moscow, Russia*

Liposomes are widely used in medicine for encapsulation of biologically active compounds. It was shown earlier that anionic liposomes with encapsulated drugs can be immobilized on the surface of the cationic chitosan nanoparticles via a direct electrostatic adsorption without additional modification of both components. This study is aimed at investigation of biodegradation of multiliposomal containers based on chitosan under the action of enzymes present in biological fluids: blood, plasma and serum (lipases, lysozyme, etc.). This will help to formulate the conditions that will ensure the removal of containers from the body after completing the drug delivery mission.
Liposomes were prepared from electroneutral phosphatidylcholine (PC) and negatively charged cardiolipin (CL). The chitosan particles were obtained from a polymer with average molecular weight of 10,000, 63,000, and 300,000 with a molar content of amino groups of 0.85. PC/CL liposomes were electrostatically adsorbed on the surface of chitosan particles. Biodegradation of liposome-chitosan complex particles was initiated by addition of a proteolytic enzyme: lipase hydrolyzed ester bonds in the satellite liposomes, or lysozyme hydrolyzed glycosidic bonds in the chitosan shell, or Morikrase, which was a mixture of enzymes capable of cleaving both ester and glycosidic bonds. The degradation was controlled by measuring particle size in the complex-enzyme suspension by dynamic light scattering. In the control experiment without enzymes, no change in the size of the complexes was observed for 96 hours. Addition of a lipase to nanoconjugate suspension led to a decrease in the size of the particles to 270 nm that is close to the size of initial chitosan before complexation with liposomes. Therefore, we can conclude that lipase selectively hydrolyzed ester bonds in lipid molecules but did not attack the chitosan core. When lysozyme was added, the size of nanoconjugates decreased to 120–150 nm, which indicated the destruction of the chitosan core. The addition of Morickrase was accompanied by a decrease in the particle size to 10–15 nm, which was significantly smaller than the size of any component of the complex: chitosan particles or liposomes. Evidently, such deep degradation of the liposome-chitosan complex might arise from the combined action of the Morikrase enzymes capable of cleaving lipids and chitosan. The above results show the enzyme-induced biodegradation of the liposome-chitosan complex to small fragments which can be easily eliminated from the human body.
Thus, the findings of this work seem to be important for the development of technologies for multiliposomal containers obtaining.

Acknowledgements. This work was supported by Russian Foundation for Basic Research (project 19-03-00314).

INTERACTION OF MULTICOMPONENT ANIONIC LIPOSOMES WITH A CATIONIC DENDRIMER

K. Trosheva[1], A. Efimova[1], Z. Shifrina[2], A. Yaroslavov[1]

[1] *Polymer Department, Faculty of Chemistry, M.V.Lomonosov Moscow State University, Moscow, Russia, ksyu.trosheva@gmail.com*
[2] *A.N. Nesmeyanov Institute of Organoelement Compounds of Russian Academy of Sciences, Moscow, Russia*

Potential medical applications of dendrimers (three-dimensional branched macromolecules of regular structure) as contrast agents for magnetic resonance imaging and means of drug delivery are being actively considered in recent decades. The advantages of dendrimers are their monodispersity and the ability to control the size and chemical composition of peripheral groups, internal levels and the nucleus. The introduction of nitrogen-containing aromatic fragments into the structure of dendrimers opens up new possibilities for their use in modern technologies. To study the physicochemical aspects of the interaction of polymers with cell membranes spherical bilayer lipid vesicles (liposomes) are used as model systems.

The purpose of this work is to study the interaction of the water-soluble cationic poly(phenylene-pyridyl) dendrimer of the third generation D366+ with anionic liquid and solid liposomes. Two-component liposomes were obtained from electroneutral phosphatidylcholine/dipalmitoylphosphatidylcholine and negatively charged cardiolipin. Three-component liposomes additionally contained cholesterol, its mass fraction ranged from 5 to 30%.

The influence of the phase state of the membrane on the nature of the interaction of liposomes with the dendrimer has been studied. It was shown that dendrimer forms complexes with all types of liposomes. The interaction is followed by the neutralization of the surface charge of liposomes, increase of the particles size and quenching of the fluorescence of a label embedded into the lipid bilayer. The interaction of the dendrimer with two-component liposomes is reversible: the polycation can be removed from the liposomal surface by adding a low molecular weight electrolyte. Dendrimer adsorption on the surface of cholesterol-containing liposomes is accompanied by the formation of defects in the membrane. It has been established that complexes between dendrimer and cholesterol-containing liposomes also could be formed in water-salt environment with a physiological salt concentration.

Acknowledgements. This work was supported by Russian Foundation for Basic Research (project 19-03-00314.

CROSS-LINKED CATIONIC STARCH SORBENTS FOR REMOVAL OF IBUPROFEN FROM WATER

V. Navikaite-Snipaitiene[1], R. Rutkaite[1], D. Simanaviciute[1], K. Almonaityte[1], V. Vaskeliene[2], R. Raisutis[2]

[1] *Department of Polymer Chemistry and Technology, Kaunas University of Technology, Kaunas, Lithuania, vesta.navikaite@ktu.lt*
[2] *Prof. Kazimieras Barsauskas Ultrasound Research Institute, Kaunas University of Technology, Kaunas, Lithuania*

The world population is ageing and the consumption of medications is constantly growing. Subsequently, the concentration of their residues in soil, surface and ground waters is also growing and might cause serious danger to human health and other living organisms. Most common pharmaceutical contaminants which could be find in the environment are antidepressants, analgesics, antiphlogistic and hormone drugs, antihypertensives and cardiovascular drugs [1]. Unfortunately, traditional mechanical and biological wastewater treatment technologies are insufficient for the removal of this type of contaminants.
The aim of this research was to study the binding of ibuprofen to cross-linked cationic starch sorbent in water.
Cross-linked cationic starch microgranular sorbent was obtained by the means of chemical and physical modification of potato starch. CCS (degree of substitution of quaternary ammonium groups was equal to 0.42) was obtained by cross-linking potato starch with 0.1 mol/AGU (anhydroglucoside unit) of epichlorohydrin (EPCH) and cationization with 2,3-epoxypropyltrimethyammonium chloride (EPTMAC) (the molar ratio AGU:EPTMAC:NaOH:H_2O=1:0.45:0.04:4). The CCS granules were additionally treated by using continuous ultrasonication in water (treatment conditions: 40 kHz, 3 W/cm^2, 20 min., 24±1 °C).
The ibuprofen (IB) was partially removed from water by adsorption onto CCS or ultrasound-treated CCS (CCS-U) granules. The Langmuir, Freundlich and Dubinin-Radushkevich adsorption models have been used to describe the equilibrium adsorption of IB onto CCS and CCS-U (Table 1).

Table 1. Adsorption models parameters for adsorption of IB on CCS and CCS-U at 30 °C temperature

Sample	Langmuir model				Freundlich model		Dubinin-Radushkevich model	
	Q_L, mmol/g	K_L, 1/mol	*EF*	R^2	n_F	R^2	E_{DR}, kJ/mol	R^2
CCS-IB	1.51	3276	0.83	0.9975	2.67	0.9796	11.11	0.9851
CCS-U-IB	1.83	2211	1.00	0.9922	4.21	0.9847	13.19	0.9878

According to the Langmuir adsorption model, the IB molecules were adsorbed on the active centers i.e. quaternary ammonium groups CCS or CCS-U. The amount of the adsorbed IB (Q_L=1.83 mmol/g) and adsorption efficiency (*EF*=1.00) were higher by 21.19 % and 20.48 %, respectively, using CCS-U microgranules as a sorbent. The higher values of Freundlich constant n_F and Dubinin–Radushkevich adsorption energy E_{DR} indicated that conditions for IB adsorption on CCS-U were more favourable and the ion-exchange mechanism was predominant during adsorption on both adsorbents.

Acknowledgements. This research was supported by the Research, Development and Innovation Fund of Kaunas University of Technology (project grant No. PP-91D/19).

References

1. W.C. Li., Environ. Pollut., **187**, 193–201 (2014).

ANTIOXIDANT ACTIVITY OF β-CAROTENE LOADED THREE-COMPONENT PARTICLES

E. Celitan, R. Gruškienė, J. Sereikaitė

Department of Chemistry and Bioengineering, Vilnius Gediminas Technical University, Vilnius, Lithuania, ruta.gruskiene@vgtu.lt

Carotenoids are a group of pigments naturally present in vegetable raw materials. They have biological properties and used mainly in food, pharmaceutical, and cosmetic industries. The "core" structural element of carotenoids is a polyene backbone consisting of a series of conjugated C=C bonds. This particular feature is primarily responsible for both their pigmenting properties and the ability of many of these compounds to interact with free radicals and singlet oxygen. Therefore, carotenoids act as effective antioxidants [1, 2]. However, during the processing and storage, carotenoids can easily undergo isomerization or oxidation reactions with the consequent decrease or loss of the colorant and biological properties. The most applied alternative to increase the stability of carotenoids and to allow their incorporation in hydrophilic environment is the microencapsulation technique, which provides a physical barrier protecting the pigment [3].

In this work, the synthesis of β-carotene and 2-hydroxypropil-β-cyclodextrin inclusion complexes was performed. Prepared complexes were additionally coated using different types of pectin: pectic acid, pectin with low or high degree of esterification (Fig. 1).

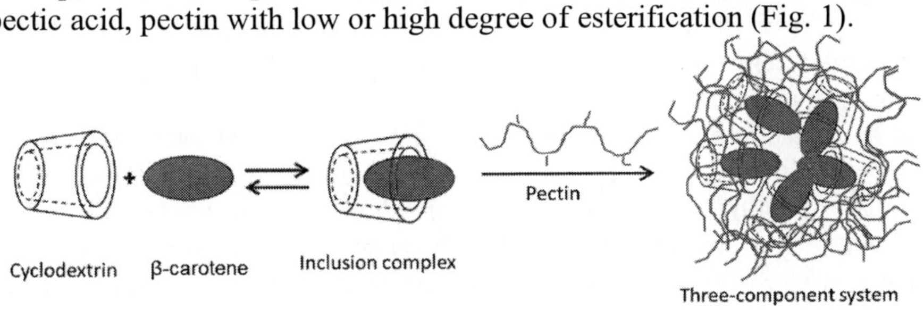

Fig. 1. Scheme of the preparation of three-component delivery system

DPPH method was used to test the ability of inclusion complex and carotene-loaded three-component particles to act as free radical scavengers and to evaluate their antioxidant activity [4]. The influence of pectin type and pH of solution on the antioxidant activity was evaluated. It was found that the pH of the pectin solution had no significant effect on the antioxidant activity, as well as the type of pectin, although high methoxyl pectin-coated particles had the slightly better antioxidant activity and its retention during storage. Based on the experimental results, it can be stated that the three-component particles retain higher antioxidant activity for longer time than the β-carotene – cyclodextrin inclusion complexes. Thus, additional coating with pectin ensures higher environmental resistance on β-carotene and helps to maintain antioxidant stability and other useful properties for longer time.

References

1. J. Young, G. L. Lowe. Antioxidants, 7, 28 (2018).
2. S. Kiokias et al., Curr. Res. Nutr. Food Sci. Jour., 4, 25-37 (2016).
3. I. L. Nunes et al., Braz. Arch. Biol. Technol., 50, 893-900 (2007).
4. Yuan et al., Carbohydr. Polym., 91, 385-389 (2013).

STRUCTURE AND THERMAL PROPERTIES OF CHITOSAN/DEEP EUTECTIC SOLVENT FILMS CONTAINING LACTIC ACID AND CHOLINE CHLORIDE

M. P. Sokolova[1,2], V. K. Vorobiov[1], A. L. Nikolaeva[1], I. V. Abalov[1], I. V. Gofman[1], A. V. Smirnov[3], M. A. Smirnov[1]

[1] *Institute of Macromolecular Compounds RAS, Saint-Petersburg, Russia, pmarip@mail.ru*
[2] *Saint Petersburg State University, Saint Petersburg, Russia*
[3] *ITMO University, Saint-Petersburg, Russia*

Chitosan (CS) is a natural polysaccharide with great potential for practical application, mainly due to its biodegradability, biocompatibility and low-toxicity. In addition, CS has good film-forming ability. However, CS cannot be extruded because its thermal decomposition is below the melting temperature. This has led to intensive elaboration of thermoplastic materials based on the CS modified either chemically or by the addition of plasticizers. Thus, searching for suitable "green" plasticizers for chitosan is an important scientific task. Deep eutectic solvents (DES) were shown to be promising plasticizer for CS [1]. DES are fluids, which are usually composed of two or more cheap and safe components that are capable for the self-association, often through hydrogen bond interactions, to form an eutectic mixture with a melting point significantly lower than one for individual component. In the case of using of aminoacids, organic acids, sugars, or choline derivatives as DES components, these liquids are meet green chemistry principles. Thus, the investigation of thermal and mechanical properties of CS/DES materials is an important task.

In this work, the eutectic mixture of choline chloride (ChCl) and lactic acid was used for the preparation of CS films by solution casting method. The content of DES in films was from 0 to 82 wt%. The impact of DES content on the morphology and crystalline structure of films was investigated using scanning electron microscopy, wide-angle and small-angle X-ray scattering. FTIR spectroscopy confirms the interactions between CS and DES components via formation of hydrogen and ion bonds. The thermal properties of the composite films were studied by simultaneous thermogravimetric and differential thermal analysis. Thermo mechanical analysis demonstrated appearance of two transitions in the temperature ranges -23 – -5 °C and 54 – 102 °C in dependent on DES content.

Acknowledgements. This work was supported by the Russian Foundation for Basic Research (RFBR), grant 18-08-01392 A. The experimental work was facilitated by the equipment of the Resource Centre of X-ray Diffraction Studies, Center for Optical and Laser materials research, Centre for Innovative Technologies of Composite Materials, Centre for Microscopy and Microanalysis and Interdisciplinary Resource center for Nanotechnology at St. Petersburg State University.

References

1. M. P. Sokolova et al., Carbohydrate Polymers, **197,** 548–557 (2018).

MECHANICAL UNFOLDING OF A UNIMOLECULAR MICELLE FORMED BY AMPHIPHILIC COMBLIKE COPOLYMER

A. S. Ivanova, A. A. Polotsky

Institute of macromolecular compounds of the Russian academy of sciences, Saint-Petersburg, Russia, sasinaas@yandex.ru

Recent advances in polymer chemistry enable the design of macromolecules with increasingly complex, well-controlled branched architecture, in particular, comblike copolymers, or graft-copolymers – polymer chains with regularly grafted side chains. This led to a novel paradigm in materials design which allows regulation of physical properties though precise control of branching topology, density of branches and their dimensions.

In the present work, we have performed a self-consistent field (SCF) modeling of mechanical unfolding of an amphiphilic comblike macromolecule in a selective solvent by using the Scheutjens-Fleer numerical SCF approach. This method allows obtaining conformational and thermodynamic properties of the considered system (e.g., the force-extension curves) in a wide range of length and grafting density of the side chains, and the quality of the solvent for the side chains.

Amphiphilic comblike macromolecule with a solvophilic backbone and solvophobic side chains can form a unimolecular micelle of spherical shape with the solvophobic core made of side chains and the soluble corona made of the loops of the backbone. In our study, it was assumed that the solvent is good (athermal) for the main chain but poor for the side chains. We demonstrate that in the case of loose grafting of side chains and more dense grafting under moderately poor solvent conditions, the stretching of the comb-like macromolecule is determined by deformation of the backbone. The force-extension curve coincides with the force–extension dependence for a freely jointed linear chain with the same degree of polymerization as the main chain of the graft copolymer.

Upon a deterioration of the solvent quality, the shape of the force-extension curve becomes more complex, and three regimes are observed, as in the case of the globule formed by a linear macromolecule: (1) At small deformations the micelle shape changes from the sphere to the prolate spheroid, the restoring force grows. (2) At moderate deformations, microphase segregation within the macromolecule occurs, the macromolecule acquires the tadpole conformation with a large globular (micellar) head and stretched tail co-existing in equilibrium, the restoring force weakly decreases. (3) At strong deformations, the micelle is completely unfolded, the restoring force grows, the deformation behavior is determined solely by the backbone stretching.

The most interesting and nontrivial scenario is observed for comblike copolymers with densely grafted long side chains under very poor solvent conditions. In this case, stretching of the micelle leads to the release of small clusters containing several side chains. The macromolecule, therefore, acquires the tadpole conformation with the tail having the pearl-necklace structure. Formation of each new cluster is accompanied by a drop in the reaction force, and the force-extension curve in this regime has a saw-tooth shape. At strong deformations, the clusters have approximately the same size. In this regime, the stretching is accompanied by the increase in. the number of clusters and, correspondingly, the decrease in the aggregation number – the number of side chains in each cluster.

Acknowledgement. This study was supported by Ministry of Education of the Russian Federation within State Contract № 14.W03.31.0022.

FORMATION AND PROPERTIES OF CAFFEIC ACID AND CHITOSAN COMPLEXES

D. Simanaviciute[1], D. Liudvinaviciute[1], R. Rutkaite[1], V. Coma[2]

[1] *Department of Polymer Chemistry and Technology, Kaunas University of Technology, Kaunas, Lithuania, deimante.simanaviciute@ktu.lt*
[2] *University of Bordeaux, UMR 5629, CNRS, LCPO, France*

Recently there has been increasing interest in phenolic acids like caffeic acid (CA) because of their antioxidant, anti-inflammatory, and anticancer activities [1]. Being hydroxycinnamic acid derivative CA has high antioxidant activity due to the presence of conjugated double bonds in the side chain, however, this makes CA vulnerable to heat, oxygen, light and moisture. Stability of phenolic acids can be improved during formation of complexes with natural polymers such as chitosan (CH). CH is soluble in acidic media, in which the protonation of amino groups occurs. After protonation D-glucosamino units of CH are positively charged, and therefore CH can form polyelectrolyte complexes with negatively charged polymers and low molecular compounds [2].
The aim of this work was to investigate the formation of CA and CH complexes (CA-CH) and to evaluate their thermal, antioxidant and water sorption properties.
To obtain CA-CH complexes (Fig. 1) CA aqueous solution of the concentration of 1 g/l was poured over CH powder, stirred with magnetic stirrer at room temperature (22 ± 1 °C) for 60 min and then the mixture was filtered. Obtained CH powder with adsorbed CA was filtered and dried at room temperature. To obtain CA-CH complexes with higher amounts of CA, procedure described above was repeated with CA-CH powder obtained in the previous step. The composition of obtained RA-CH powders was evaluated as the mole ratio of CA to CH (mol/mol).
CA-CH complexes with CA to CH molar ratio of 0.05 mol/mol, 0.16 mol/mol, 0.46 mol/mol and 0.54 mol/mol were obtained. FT-IR spectroscopy was used to confirm the formation of CA-CH complexes.

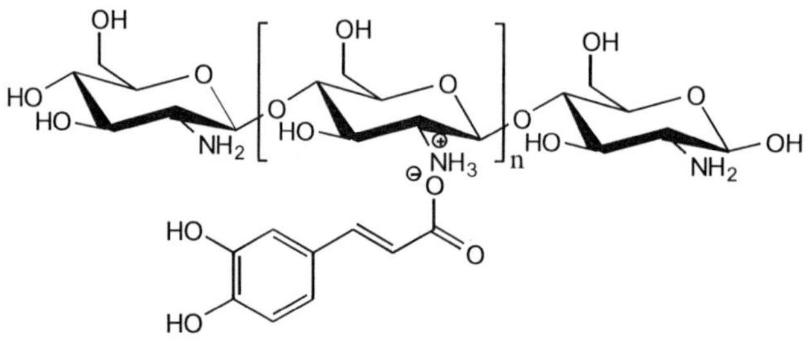

Fig. 1. Structure of CA-CH complexes

Thermal degradation of CA, CH and CA-CH complexes was determined by TG analysis. Thermal destruction of CA-CH complexes occurred at lower temperatures than that of CH or CA.
Meanwhile, water vapor sorption and antioxidant activity of CA-CH powders depended on molar ratio of CA-CH complexes. Water vapor sorption decreased and antioxidant activity increased with increased amount of CA in the complex.

Acknowledgements. The financial support of the Research Council of Lithuania for the Lithuanian-French programme "Gilibert" project No. S-LZ-19-6 is highly acknowledged.

References

1. F. Natella et al., J. Agric. Food Chem., **47**, 1453–1459 (1999).
2. Y. Luo et al., Int. J. Biol. Macromol., **64**, 353–367 (2014).

STABILIZATION OF COLLAGEN STRUCTURE WITH MONTMORILLONITE DISPERSIONS

M. P. Zhaldak[1], **O. R. Mokrousova**[1]

[1] *Department of Commodity Science and Customs Affairs, Kiev National University of Trade and Economics, Kiev, maryna070992@ukr.net*

A collagen is the basic protein of derma in the skin of animals. The presence in the structure of the protein of the carboxyl, amine, peptide and others groups assists effective stabilization of the collagen structure of derma. During the processing of animal skins into natural leather there is transformation of bonds in a collagen. Thus, chemical materials cooperate with functional groups protein and the formation new chemical bonds. It assists stabilizing of structure of derma. As a result, a volume capillary-porous structure is formed with the necessary level of operating and hygienical properties in leather.

The tanning process plays an important value stabilizing of collagen structure of derma. Today different mineral and organic tanning agents are used. The basic chromium sulfate is used as a mineral tanning agent to produce around 80% of the leather in the world. It is a complex compound that effectively forms the structure of the derma. The main stabilization of the collagen structure is through the formation of coordination bonds with the carboxyl groups of the protein The disadvantage of chrome tanning is inefficient use of chromium compounds, as almost 40% of them remains in the waste tanning solution [1-2].

For the ecologization of the production of leather, use of natural clay minerals. Known information [3] indicates the possible stabilizing derma collagen structure with dispersions of montmorillonite, which have been modified by chromium compounds.

In this paper, the stabilization of collagen structure is analysed by dispersions of montmorillonite, treated with aluminum compounds.

Modified dispersions of montmorillonite and technical gelatin (as a model of derma collagen) is used for the study. Prepared films, which are then analyzed on on the universal Fourier transform infrared spectrometer TENSOR-37 (BRUKER, Germany) within the 4000 - 400 cm^{-1}.

The IR spectroscopic analysis revealed an effective physicochemical interaction between collagen and modified dispersions of montmorillonite. It is confirmed by the formation of numerical bonds involving the functional groups of gelatin and the active centers of the mineral. The largest changes are observed in a high-frequency and low-frequency ranges. Accordingly, spectrums characterize the valency fluctuations. that participate in the formation of hydrogen connections. The changes of bands of fluctuations. of carboxyl and amine groups of gelatin are set with the hydroxocomplex ions Cr (III) and Al (III).

The presented results of chemical interactions indicate the formation of hydrogen, ionic and covalent bonds, which may affect the stabilization of the structure of the derma during tanning process.

References

1. Z. Zhongkai Physicochemical Properties of Collagen, Gelatin and Collagen Hydrolysate from Bovine Limed Split Wasters *JSLTC,* **1(90),** 23-29 (2006).
2. C. Gaidau Applicative Chemistry of Tanning Metallic Heterocomplexes *Benthan Science Publishers*, 130 p. (2013).
3. M. Marukhlenko, V. Palamar, O. Mokrousova Stabilizing derma collagen structure with modified dispersions of montmorillonite *Materials Science and Engineering*, **1(111),** 1-8 (2016).

PHOTO-TRIGGERED CAPSULES

D. Pirone[1,2,3], V. Marturano[4], R. Del Pezzo[1,2,3], S. Fernandez Prieto[3], T. Underiner[5], M. Giamberini[1] and B. Tylkowski[1,2,5,*]

[1] *Department of Chemical Engineering, Rovira i Virgili University, Av. Països Catalans 26, 43007 Tarragona*
[2] *Centre Tecnològic de la Química de Catalunya, Carrer Marcelli Domingo s/n, 43007 Tarragona, Spain*
[3] *The Procter and Gamble Company, Temselaan 100, Strombeek-Bever 1853, Belgium*
[4] *Department of Chemical, Materials, and Production Engineering (DICMAPI), University of Naples "Federico II", P. le Tecchio, 80, 80125 Napoli, Italy*
[5] *The Procter and Gamble Company, 6210 Center Hill Avenue, Cincinnati, OH 45224, USA*

The development of photo-responsive capsules to tune and control the sustained-release of encapsulated actives is a fascinating and challenging route to improve the performances and effectiveness of a wide range of delivery applications. 1-3 In this work, we report the preparation of visible light-responsive capsules obtained *via* oil-in-water interfacial polycondensation between modified diacyl-chloride azobenzene moiety and diamine flexible spacer in the presence of cross-linkers with different structures and functionalities. The effect on the release profile of the encapsulated perfume oil was investigated using three flexible spacers with different lengths (1,8-diaminooctane; 1,6-diaminohexane and 1,4-diaminobutane) and two types of cross-linkers (1,3,5-benzenetricarbonyl trichloride and melamine). We analyzed how the microcapsules properties can be tailored changing the design of the shell structure. Fine tuning of the perfume release profiles was obtained. The changes in capsules size and morphology due to visible light irradiation were monitored via light scattering, optical microscopy and atomic force microscopy. Perfume release was 50% faster in the systems prepared with melamine as the cross-linker. Modelling studies were carried out to support the discussion of the experimental results.

Acknowledgements. Financial support from European Community's Seventh Framework Programme (FP/2007–2013) under IOF Marie Curie grant agreement no. 328794, Horizon2020 ITN Marie Curie grant agreement no. 675624 are gratefully acknowledged.

References

1. Marturano, V., Cerruti, P., Giamberini, M., Tylkowski, B. & Ambrogi, V. Light-Responsive Polymer Micro- and Nano-Capsules. *Polymers* **9**, 8 (2017).
2. Marturano, V. *et al.* Essential oils as solvents and core materials for the preparation of photo-responsive polymer nanocapsules. *Nano Res.* **11**, 2783–2795 (2018).
3. Ping, Y. *et al.* pH-Responsive Capsules Engineered from Metal-Phenolic Networks for Anticancer Drug Delivery. *Small Weinh. Bergstr. Ger.* **11**, 2032–2036 (2015).

CHITOOLOGOSACCARIDE AND ITS DERIVATIVES: SYTHESIS, INVESTIGATION AND DETERMINATION OF ANTIBACTERIAL ACTIVITY

A. Bočkuvienė[1], L. Juravičius[1], R. Stanevičienė[2], E. Servienė[2]

[1] *Institute of Chemistry, Vilnius University, Vilnius, Lithuania, alma.bockuviene@chf.vu.lt*
[2] *Nature Research Centre, Vilnius, Lithuania.*

Chitooligosaccharides (CHIOS) are low molecular weight compounds produced from high molecular weight chitosan by several methods such as enzymatic and acidic hydrolysis. The molecular weights of CHIOS are 10 kDa or less, they are water soluble and stable in all pH range [1, 2]. Moreover, chitosan itself is a non-toxic and biodegradable biopolymer [3]. CHIOS and its derivatives are remarkable biomaterials because of its numerous biological and immunological properties: antioxidant, anti-allergic, anti-inflammatory, anticoagulant, anti-cancer, anti-bacterial, anti-human immune deficiency virus, anti-hypertensive, anti-Alzheimer's, anti-diabetic and anti-obesity [4].
CHIOS was prepared by acidic hydrolysis in microwave reactor and modified by aldehydes: octanal, butanal, methanal and chlorsulfonic acid to obtain N-alkylated, sulfo- and N-alkylsulfochitooligosaccharides. CHIOS and its derivatives were analyzed by acid-base titration, potentiometric, conductometric, SEC, DLS methods; structure was evaluated by FT-IR, ^{1}H-NMR techniques. The antibacterial activity was investigated against the growth of *E. coli* and *B. subtilis* bacteria.
High molecular weight chitosan was hydrolyzed and obtained 7.6 kDa of water-soluble CHIOS, with degree of deacetylation up to 60%. The yield of CHIOS and its derivatives ranged from 25% to 67%. N-alkylated derivatives degree of cationization varied from 26% to 60% and degree of cationization increased with increasing aldehyde group length. Degrees of sulphonation of sulfochitoligosaccharides and N-alkylsulfooligosaccharides ranged from 0.62 to 1.78. Conductometric titration revealed that N, O, O-substituted derivatives were synthesized. DLS measurements showed that the hydrodynamic radius of the N-alkylated derivatives decreases from 31.57 nm to 16.28 nm. The size of the particles of N-alkylsulphochitooligosaccharides are greater than CHIOS.
The highest antibacterial activity against *B. subtilis* was observed for CHIOS and N, N-dimethyl-N-octylchitooligosaccharide with minimum inhibitory concentration ranging from 125 to 250 μg/mL. The highest inhibition of *E. coli* bacteria was for CHIOS and minimum inhibitory concentration for both exponential and stationary cultures was 625 μg/mL approximately. The results of antibacterial activity have shown that CHIOS and its derivatives are potential bio-materials that can be used in food, cosmetics and medicine fields.

References

1. M. Rinaudo, Prog. Polym. Sci.,**31**, 603–632 (2006).
2. B. B. Aam et al., Mar. Drugs, **8**, 1482-1517 (2010).
3. V. K. Mourya et al., Polym. Scie., **53**, 583–612 (2011).
4. F. Liaqat,R. Eltem, Carbohydr. Poly., **184**, 243–25 (2018).

POLYVINYL BUTYRAL FILMS WITH ANTIMICROBIAL ACTIVITY

V. Jankauskaitė[1], I. Danisevičienė[1], A. Andziukevičiūtė-Jankūnienė[1], V. Raudonienė[2], A. Paškevičius[2]

[1] *Department of Production Engineering, Kaunas University of Technology, Kaunas, Lithuania, virjan@ktu.lt*
[2] *Nature Research Center, Institute of Botany, Vilnius, Lithuania*

Airborne microbes are biological contaminants like bacteria, viruses or fungi as well as airborne toxins passed from one victim to the next through the air, without physical contact. It is believed that about one third of indoor air quality problems are microbe-related [1]. Antimicrobial coatings are rapidly emerging for preventing the bacterial colonization and limiting the spread of infections [2].

The scope of the current research was to assess the feasibility of coating air recuperation system surfaces with thin biocidal polyvinyl butyral (PVB) films. PVB was chosen due to the optical clarity, adhesion to many surfaces, toughness and flexibility. Biocides with very broad antimicrobial spectrum of activity, such as triclosan, pyrithiones and sodium or copper pentachlorophenate, were solution blended with PVB composition [3].

The PVB films of various compositions were analyzed for their structure and thermal properties. Antimicrobial properties of films were determined by Kirby-Bauer disk diffusion susceptibility test. It was determined that biocidal PVB films show high inhibition activity to various pathogens, such as fungus (*Aspergillus spp., Candida spp., Penicillium brevicompactum, Scopulariopsis brevicaulis, Cladosporium cladosporioides*), bacteria (*Escherichia coli,* MSSA, MRSA) and nosocomial pathogens (*Pseudomonas aeruginosa, Acinetobacter baumannii*) (Fig. 1).

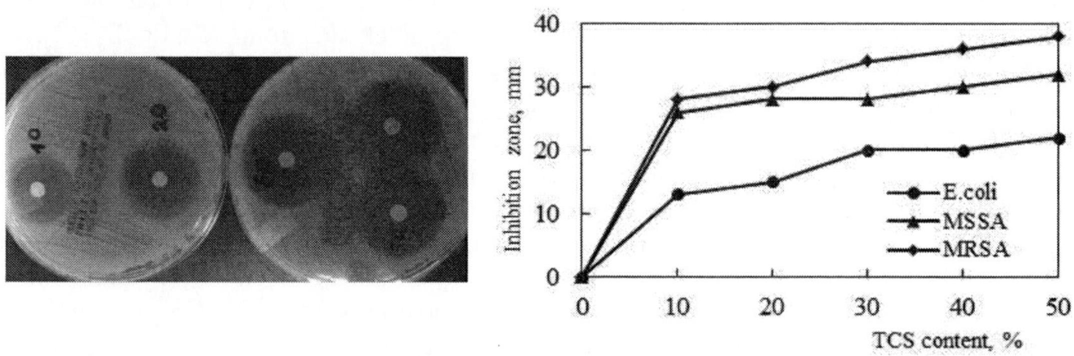

a b

Fig. 1. Antibacterial activity of PVB coatings: a – inhibition zone against MRSA by addition copper pentachlorophenate; b – inhibition zone against different bacteria strains by addition triclosan (TCS).

The spraying of low viscosity PVB compositions were utilized to obtain biocidal coatings on different surfaces (aluminum, etc.). PVB coated metallic surface was examined regarding changes in topography.

Acknowledgements. This research was funded by the European Social Fund and Republic of Lithuania grant number S-J105-LVPA-K-01-0152.

References

1. Norhidayah et al., Procedia Engineer., **53**, 93–98 (2013).
2. M. Cloutier et al. Trends Biotechnol., **33**, 637–652 (2015).
3. W. Bojar et al., Adv. Clin. Exp. Med., **18**, 401–405 (2009).

EFFECT OF CATIONIC BRUSH COPOLYMERS ON COLLOIDAL STABILITY OF $GdPO_4$ PARTICLES WITH DIFFERENT MORPHOLOGY

V. Klimkevičius, A. Babičeva, M. Janulevičius, R. Makuška and A. Katelnikovas

Institute of Chemistry, Vilnius University, Naugarduko 24, LT-03225 Vilnius, Lithuania, vaidas.klimkevicius@chf.vu.lt

Development of both micro and nano sized particles with different structure and morphology has increased significantly in recent years as morphology could be responsible for some of physicochemical properties of particles. Rare-earth doped –micro and –nano orthophosphate particles are promising host materials as they possess good both chemical and thermal stability, appropriate luminescent properties and are already applied as luminescent phosphors, down/up-conversion materials, catalysts [1, 2]. As phosphates are biologically inert host – these particles show high potential to be applied in bio-related fields such as MRI contrast agents, bio-labeling, drug delivery etc. [3]. However, applicability of such particles is limited by poor stability in aqueous solutions. The usage of various commercial surfactants (e.g. oleic acid, TRITON, TWEEN) cannot ensure desired stability of $LnPO_4$ particles. It is known that polymeric brush electrolytes have high potential to surface modification and stabilization of other nanoparticles (e.g. SiO_2, TiO_2, Al_2O_3) [4, 5] and in our case the usage of non-linear polymeric materials could be a solution.

In this study, we present the synthesis of cationic brush polyelectrolytes and their use in stabilization of $GdPO_4$ particles in aqueous media. Polymers of various compositions were synthesized via RAFT polymerization route. SEC equipped with triple detection (RI, DP, RALS, LALS) were used to determined molecular parameters (M_n, M_w, M_w/M_n). Thus exact composition of synthesized polymers were determined using NMR spectroscopy. Cationic brush polymers were used to improve stability of aqueous $GaPO_4$ particle dispersions. Firstly, the IEP points of different morphology $GdPO_4$ particles (nanorods, nanoprisms, nanospheres) were found by measuring zeta potential of bare particle dispersions under various pH values. The oppositely charged particles (negatively charge in alkaline media pH ~ 10) in dispersion were treated by cationic polyelectrolytes of different compositions. The concentration and composition effects of used polymers on change of particle surface potential and stability (DLS measurements) in dispersions were investigated and presented in this work.

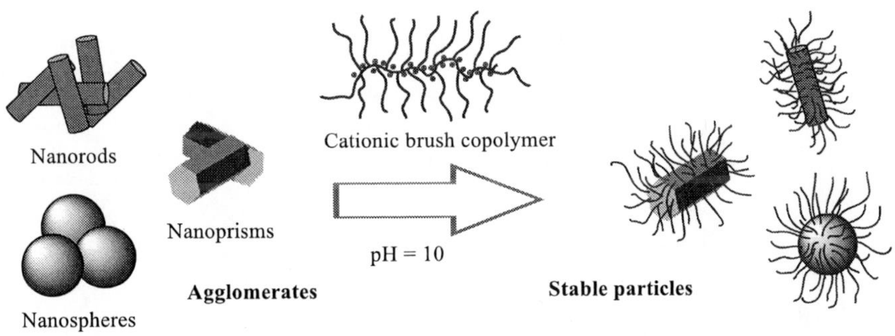

Fig. 1. Graphical illustration of $GdPO_4$ with different morphology (nanorods, nanoprisms, nanosperes) stabilization by using cationic brush copolymers

References
1. Z. Wang et al., CrystEngComm, **20** (2018) 796-806.
2. M. Ferhi et al., Journal of Alloys and Compounds **714** (2017) 144-153.
3. Q. Yang et al., RSC Advances, **8** (2018) 12832-12840.
4. V. Klimkevicius et al., Langmuir **31** (2015) 2074–2083.
5. Y. de Hazan et al., Journal of Colloid and Interface Science **365** (2012) 163–171.

A STUDY ON PROPERTIES OF CERIUM DOPED YTTRIUM ALUMINIUM GARNET AND POLYMER COMPOSITES

G. Inkrataitė[1], J. Aglinskaitė[2], R. Skaudžius[1], P. Vitta[2]

[1] *Institute of Chemistry, Vilnius University, Naugarduko 24, Vilnius, Lithuania, greta.inkrataite@stud.chgf.vu.lt*
[2] *Institute of Photonics and Nanotechnology, Vilnius University, Saulėtekio av. 3, Vilnius, Lithuania*

Cerium doped yttrium aluminum garnet is one of the most frequently used fluorescent materials. First studies on this material and its applications have been done as earl as year 1967 [1]. However, in the recent years research on this garnet has been revitalized, because of its use in the new generation of LED lighting. Even so, one of the main drawbacks of diodes is the problem of their lackluster lifespan. In order to improve longevity of the LEDs as well as the intensity of the emitted light, new research is focused on making new YAG:Ce composite materials as well as their integration into the diode itself [2]. Over the last few decades scientist have been trying to improve the luminescence intensity of phosphors. One way to do so is to use laser as an excitation source. However, this makes phosphors heat up, and in order to prevent that measures need to be taken. Sometimes boron nitride is added together with the phosphor in order to improve thermal conductivity. Also in literature it is also reported that luminescence properties can be improved by addition of certain polymers [3].
For this project different composites were prepared, which could be used in LED in order to improve their longevity and reliability. For the reason, YAG:Ce phosphor powder was synthesized by sol-gel combustion method and then mixed together with differing amounts of BN and either M600 or M280 monomer which were polymerized under UV light. Composites were analyzed by x-ray diffraction (XRD), scanning electron microscopy (SEM) and differential scanning calorimetry (DSC). Also, luminescence properties such as quantum efficiency, decay times, emission spectrum.

Acknowledgements:
This research is funded by the european social fund under the no 09.3.3- lmt k-712 " development of competences of scientists, other researchers and students through practical research activities" measure.

References

1. V. Bachmann, et al., Chemistry of Materials,. 21(**10**), 2077-2084, 2009.
2. R. Zhang, et al., Laser & Photonics Reviews, 8(**1**), 158-164, 2014.
3. X. Wang, et al., International Journal of Applied Ceramic Technology, 10(**4**), 610-616, 2012.

ENZYMES IN POLYMER CHEMISTRY: PRODUCTION OF BIOPOLYOLS VIA CHEMO-ENZYMATIC ROUTE

M. Šulcienė, I. Matijošytė[1], B. Kolvenbach[2]

[1] *Vilnius University, Life Sciences Center, Institute of Biotechnology, Sector of Applied Biocatalysis, inga.matijosyte@bti.vu.lt*
[2] *Institute for Ecopreneurship, School of Life Sciences, University of Applied Sciences Northwestern Switzerland*

Vegetable and other non-fossil oils are convincing starting products for plastic and polymer applications as they are inherently sustainable, renewable and biodegradable raw materials [1, 2]. They provide precursors such as polyols, which have a profound application potential for the production of polyurethane. However, research on the production of polyols based on non-fossil oils is exclusively focused on the conventional chemical synthesis pathways, and currently, no polyols are produced by enzymatic or chemo-enzymatic processes at an industrial scale. Recently, we have presented a study on a chemo-enzymatic route for biopolyol synthesis from rapeseed oil by a two-stage process: epoxidation by means of commercially available enzyme lipase Novozyme 435 and *in situ* generated peracid from linoleic acid followed by hydrolysis applying a reusable catalyst, i.e. the ion exchange resin Amberlyst-15 [3].

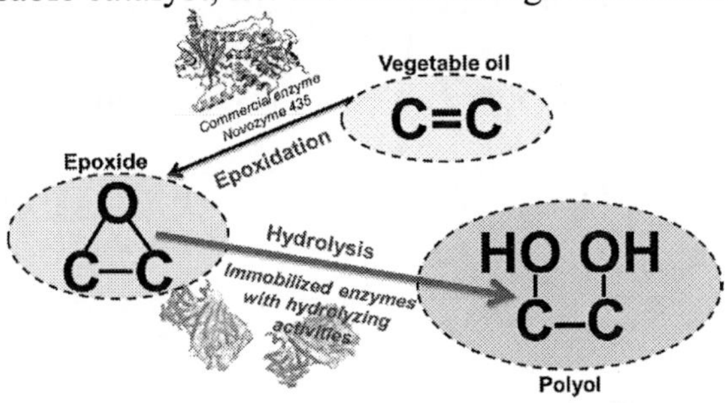

Fig. 1. The Application of enzymes with epoxidizing and hydrolyzing for biopolyols production

By this study, we were aiming to explore alternatives to commercially available enzymes immobilized onto carriers, usually which are affected by organic solvents. Hence, we investigated an effective hydrolytic route from epoxidized oils to polyols by applying enzymes immobilized onto fumed silica nanoparticles and those immobilized by the formation of CLEA. Using immobilized enzymes allows for an easy recovery and subsequent reuse of enzymes, thus reducing the costs of enzymatic production of polyol. Additionally, the use of the most widely applied cross-linkers, i.e. glutaraldehyde, p-benzoquinone (BQ) and a trifunctional derivative of acrylamide, 1,3,5-triacryloylhexahydro-1,3,5-triazine (TAT) were studied [4, 5]. The latter has, to the best of our knowledge, never before been investigated for its suitability in CLEA preparation of biocatalysts.

Acknowledgements. This research was performed during Scientific Exchange Programme „Sciex-NMSch" between Switzerland and Lithuania at FHNW.

References

1. Sh. Miao et al., Acta Biomater., **10**, 1692–1704 (2014).
2. M. Mosiewicki et al., Eur. Polym. J., **49**, 1243–1256 (2013).
3. M. Šulcienė et al., Int. J. Bio.l Macromol., **116**, 1049–1055 (2018).
4. O. Barbosa et al., RSC Adv. **4**, 1583–1600 (2014).
5. G. Dienys et al.,Chem. **9**, 744–748 (1998).

THERMOELECTRIC PROPERTIES OF PEDOT:PSS AND ANTIMONY TELLURIDE MODIFIED CARBON NANOTUBE COMPOSITES

J. Bitenieks[1], K. Buks[2], R Merijs Meri[1], J. Zicans[1], T. Ivanova[1], J. Andzane[2]

[1] *Institute of Polymer Materials, Riga Technical University, Riga, Latvia, juris.bitenieks@rtu.lv*
[2] *Institute of Chemical Physics, University of Latvia, Latvia*

Poly(3,4-ethylenedioxythiophene) polystyrene sulfonate (PEDOT:PSS) is one of the most promising electro conductive polymers that can be used for thermoelectric (TE) materials. As in a liquid form this material is beneficial for preparation of TE nanocomposites with enhanced TE properties by adding conductive and inorganic TE fillers thus improving its potential. By now PEDOT:PSS fillers include carbon nanotubes and nano-sized inorganic TE particles [1,2]. In this work PEDOT:PSS was modified with multi-walled carbon nanotubes (MWCNTs) to improve Seebeck coefficient and further improvements were made by modification with Sb_2Te_3 synthesized on MWCNT surface. All fillers at different weight concentrations were mixed in PEDOT:PSS by using ultrasonic probe and samples were obtained by drop coating method on a 25×25 mm glass surface.
Seebeck coefficient (S) measurements (Fig. 1) for unmodified PEDOT:PSS showed S=11.3 μV/K. Addition of MWCNTs from 1 to 10 wt. % showed only slight increase in S value to 12.5 μV/K at 10 wt. % concentration. Hovewer Sb_2Te_3 modified MWCNT filler allowed to increase Seebeck value to 18.5 μV/K.

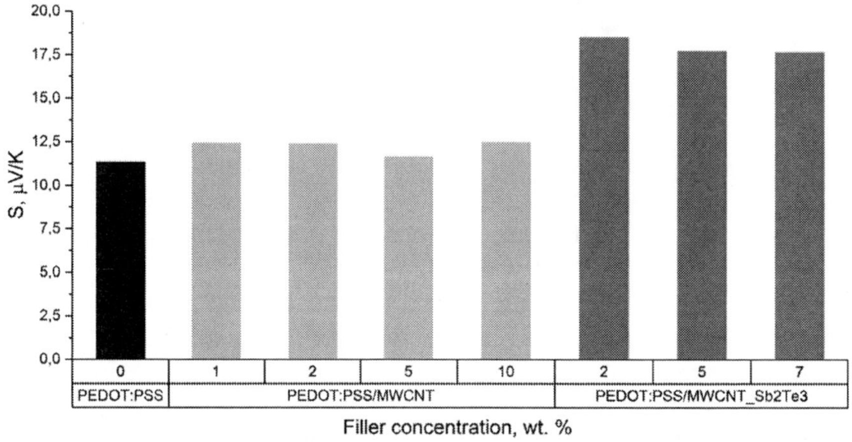

Fig. 1. Seebeck coefficient of PEDOT:PSS and PEDOT:PSS-based TE nanocomposites

Obtained results suggest, that PEDOT:PSS/MWCNT_Sb_2Te_3 nanocomposites could be used as a p type TE material for TE device manufacturing by replacing inorganic Bi_2Te_3 TE materials with TE polymer composites.

Acknowledgements. Financial support in the framework of the ERDF project Nr. 1.1.1.1/16/A/257 is greatly acknowledged.

References
1. N. Toshima, Synth. Met., **225**, 3-21 (2017).
2. W. Zheng et al., Appl. Phys. Lett., **105**, 1-4 (2014).

EVOLUTION OF POLYPROPYLENE COMPOSITES PROPERTIES DUE TO GRAIN HUSKS PREPARATION TECHNOLOGY

I. Bochkov[1], M. Varkale[1], R. Merijs Meri[1], J. Zicans[1], A. K. Bledzki[2]

[1] *Institute of Polymer Materials, Riga Technical University, Riga, Latvia, Ivans.bockovs@rtu.lv*
[2] *Institute of Materials Science, West Pomeranian University of Technology Szczecin, Poland*

Bio-based polymer composites are widely used materials in many sectors of economy in Europe. Nowadays, thanks to the "green thinking", these composites are on great interest to scientists and industry representatives. Main idea of biomass fiber usage in polymer composites is based on reduction of primary polymer application as well as integrated management of wood processing residues (wood chips, saw dust etc.). Hitherto some commercially viable examples of wood fiber as reinforcement material for polymer composites that are present on the market are produced from raw wood material from sustainably managed forests. Another biomass material that has attracted interest is grain husks. Grain husks is unavoidable by-product of grain milling. It is a "valuable" material with high cellulose content, yet there are no examples on viable industrial application of this material for now, except of incineration for energy generation. Due to a compilation of the scientific literature it has been found that husks fiber effect on composite material could be as good as wood fiber.

Blending new composites on polypropylene base can reduce main disadvantages of husk fibers such as water sorption, swelling, form instability etc. Main interest from the part of industry could be raw husks at minimal costs. Raw husks have impurities and other admixtures such as grains, leafs, seeds, stems, sand, traces of pesticides and so on. Main idea of this work is to compare different types of grain husk preparation methods to ensure manufacturing of economically attractive polymer composites.

3 different techniques of grain husks preparation have been compared: sifted grain husks – husks with separated dust fraction (1), shredded grain husks using cutting mill with the sieve of 0.25 mm (2), shredded and sieved grain husks using cutting mill with the sieve of 0.25 mm and pneumatic sieves ensuring the sieved material of sizes 0.05-0.5 mm. Effect of these 3 techniques of spelt and oat husks fiber preparation on polypropylene matrix composite properties were investigated. All husks were dried for 19 h before introduction in the polymer composite (40 wt % of husks) via twin screw co-rotating extruder at thermal profile 175-180-180-190-200° C. Testing samples were prepared by injection molding and compression molding, as appropriate. Mechanical properties and structure of the composites were investigated.

Acknowledgements. This work has been supported by European Council HORIZON 2020 Program ERA.NET Cofund project "High performance short fibre biobased hybrid composites for injection moulding" (HyBiCo).

IMPACT STRENGTH ANALYSIS OF PET/SOFT WOOD FIBRES HYBRID COMPOSITES BY DROP WEIGHT TEST

I. Bochkov[1], G. Japins[2], A. Kovalovs[2], P. Franciszczak[2]

[1] *Institute of Polymer Materials, Riga Technical University, Riga, Latvia*
[2] *Institute of Materials and Structures, Riga Technical University, Riga, Latvia*
[3] *Institute of Materials Science, West Pomeranian University of Technology Szczecin, Szczecin, Poland*

Impact test determines the amount of energy absorbed by a material during fracture. This energy may be used to determine the impact strength or impact resistance of the material depending on the test that was performed and the characteristic that is to be determined. These values are important for the selection of materials that will be used in applications. Impact test of matrix material allows comparing quantitatively the relative values of the damage resistance parameters for composite materials with different volumetric filling ratios.

The main purpose of this study was an investigation and comparison of the influence of PET (Polyethylene Terephthalate) fibres and Soft wood fibres (WEHO 500), as well as their hybrid combination on the impact strength of the reinforced composites based on polypropylene matrix. For this purpose injection moulding grade polypropylene HP400R was applied as a matrix polymer. The manufactured composite compounds were subsequently injection moulded into test specimens.

Impact strength of manufactured test specimens was measured according to ISO 6603 and using a testing machine INSTRON Dynatup 9250 HV Impact Tower. The dimensions of specimens are 50 mm × 50 mm with 2 mm thickness. Drop weight impact test took place at 50% relative humidity and 23°C. The specimens had been conditioned under the same circumstances for at least 24 h before the testing was performed. Ten specimens were tested for each combination of PET fibres and Soft wood fibres.

The experimental results revealed that adding soft wood fibers into the hybrid composites shows a slight change in energy absorption, and decreased deflection until 20%.

Acknowledgements. The study was done partially within the framework of the projects: 'Investigation on conditions for occurrence of positive hybrid effect in thermoplastic composites' 2014/15/N/ST8/03174 funded by the National Science Centre of Poland (NCN) and 'High performance short-fibre biobased hybrid composites for injection moulding (HyBiCo)' supported by the European Social Funded m-era.net network project.

INFLUENCE OF THE PET FIBRES AND SOFTWOOD FIBRES AND THEIR HYBRID COMBINATION ON THE MECHANICAL CHARACTERISTIC BY RESPONSE SURFACE METHODOLOGY

A. Kovalovs[1], K. Kalnins[1], P. Franciszczak[2]

[1] Institute of Materials and Structures, Riga Technical University, Riga, Latvia, andrejs.kovalovs@rtu.lv
[2] Institute of Materials Science, West Pomeranian University of Technology Szczecin, Szczecin, Poland

The main purpose of this study was an investigation and comparison of the influence of PET (Polyethylene Terephthalate) fibres and Softwood fibres (WEHO 500), as well as their hybrid combination on the tensile and flexural characteristics of the reinforced composites based on polypropylene matrix. For this purpose injection moulding grade polypropylene HP400R was applied as a matrix polymer. The manufactured composite compounds were subsequently injection moulded into test specimens.

The problem has been solved by the method of planning of the experiment and response surface method. The solution of the problem is divided into several stages: the choice of design parameters and the interval of variation of each parameter, the construction of the experiment plan for the selected parameters, the execution of the experiment and the determination of the general form of the regression equation according to the results of experiments.

The experiments were designed according to Full Factorial design (FFD). FFD of experiment is the most popular designs owing to their simplicity and relatively low cost. It is very useful for preliminary studies or in initial optimization steps. The plan of experiments is formulated for 2 design variables, namely, Softwood fibres (WEHO 500) and PET fibres, and 9 experiment points. Subsequently in the points of plan of experiment the results of testing were obtained. In the next stage the numerical data obtained by the testing in the points of plan of experiments was used in order to build the approximating functions. After selection of equation of regression the parametric studies were carried out to scrutinize the influence of design parameters on Mechanical properties such as tensile and flexural strength, tensile and flexural modulus.

Acknowledgements. The study was done partially within the framework of the projects:'Investigation on conditions for occurrence of positive hybrid effect in thermoplastic composites' 2014/15/N/ST8/03174 funded by the National Science Centre of Poland (NCN) and 'High performance short-fibre biobased hybrid composites for injection moulding (HyBiCo)' supported by the European Social Funded m-era.net network project No 4297.

VISCOELASTICITY OF BIO-BASED WOOD MIMIC POLYBUTYLENE SUCCINATE COMPOSITES WITH DIFFERENT CELLULOSE FILLERS AND CELLULOSE FUNCTIONALIZATION STRATEGIES

A. Barkane[1], O. Platnieks[1], S. Gaidukovs[1], G. Gaidukova[1], I. Filipova[2], M. Laka[2], M. Skute[2], V. Fridrihsone[2]

[1] *Institute of Polymer Materials, Faculty of Materials Science and Applied Chemistry, Riga Technical University, Riga, Latvia*
[2] *Latvian State Institute of Wood Chemistry, Riga, Latvia*

The viscoelasticity properties of wood mimic polybutylene succinate composite with different cellulose fillers are studied by application of the thermomechanical analysis (DMA, TMA). Composite formulations were prepared changing cellulose type and content in the compositions. Several cellulose sources of different size (microcellulose, nanocellulose) were added to the resin formulations to amend the properties. These composites were melt blended by internal kinetic mixer. The processing characteristics are temperature 120 °C, speed 60 rpm, time 6 min. While thermomechanical properties were determined as a function of cellulose fillers content and functionalization type. Cellulose filler concentration varied in the range 1-70 wt.%; cellulose surface were functionalized with silanes, polyethylene glycol, palmitic acid, isocyanate. DMA and TMA analysis for the prepared biopolymer composite's compressed films were investigated. Storage modulus, loss modulus and tanΔ were very sensitive related to the composition and cellulose surface treatment. Viscoelastic properties (storage and loss moduli, damping) of the composites were calculated from the obtained experimental data. TMA analysis testified the remarkable changes in the glass transition of the polymer after the introduction of cellulose filler. Thermal expansion coefficient was calculated from the dilatometric curves' slopes. The measured characteristics were directly dependent from the composition of the composites and the operated temperature. It resulted in improved thermal and viscoelastic properties of composite materials due to the strong interaction between the components after the functionalization.

Acknowledgements. "This research is funded by the Latvian Council of Science, project WoodMimic, No. lzp-2018/1-0136"

CHARACTERIZATION OF ELECTROSPUN NANOFIBERS OF POLY(BUTYLENE SUCCINATE) BLENDS WITH NANOFIBRILLATED CELLULOSE

O. Platnieks[1], N. Neibolts[1], S. Gaidukovs[1], A. Barkane[1], V. K. Thakur[2], I. Filipova[3], V. Fridrihsone[3], M. Enachescu[4], Z. Zelca[5]

[1] *Faculty of Material Science and Applied Chemistry, Institute of Polymer Materials, Riga Technical University, Latvia, oplatnieks@gmail.com*
[2] *School of Aerospace, Transport and Manufacturing, Cranfield University, United Kingdom*
[3] *Latvian State Institute of Wood Chemistry, Latvia*
[4] *Center for Surface Science and Nanotechnology, University Politehnica of Bucharest, Romania*
[5] *Institute of Design Technologies, Riga Technical University, Latvia*

Poly(butylene succinate) (PBS) is commercially available aliphatic polyester that can be fully bio-based and has properties like biodegradability, melt processability, and good mechanical and thermal properties[1], which make if perfectly suitable for petroleum-based polymer replacement and environment pollution reduction.
Cellulose is abundant biopolymer with unique properties that can be used to improve composite structures. Nanofibrillated cellulose (NFC) is relatively cheap and can be implemented in nanostructures, to increase properties like surface area.
PBS and NFC were electrospun to prepare nanosized fibres in nonwoven scaffolds. PBS fibres were obtained from solvent systems chloroform (CF)/methanol (MeOH) (4/1, w/w), which was selected as most suitable[1]. NFC was coated with PEG to enhance compatibility with PBS and filler content was selected at 0.5 and 1 wt%. PBS fibres average diameters were in the range of 50 -200 nm. Fibres structure and composition were investigated and characterized by SEM, DSC, FTIR, Raman spectroscopy and TGA.

Acknowledgements.
This research is funded by the Latvian Council of Science, project WoodMimic, No. lzp-2018/1-0136. The authors would like to acknowledge the COST Action CA15107 (MultiComp) for financial support.

References
1. W. Klairutsamee et al., J Appl Polym Sci., **132,** 43 (2015).

THE EFFECT OF FIXED NITROGEN CONTENT IN MODIFIED PLYWOOD DUST AS A FILLER ON PROPERTIES OF WOOD-POLYMER COMPOSITES

G. Shulga, B. Neiberte, A. Verovkins, J. Jaunslavietis, S. Vitolina, T. Betkers

Latvian State Institute of Wood Chemistry, Riga, Latvia, shulga@junik.lv

Waste polypropylene, the polymeric municipal solid waste, can be used as a polymer matrix in the developed wood-polymer composites (WPCs). The joint usage of recycled polypropylene together with a filler from waste lignocellulosic materials is an advanced practice for both obtaining WPCs and improving the environment. At the same time, the studies dedicated to WPCs based on recycled polypropylene are limited, and the impact of recycled polypropylene on the WPCs properties is still not fully understood and represents the new research opportunities. The aim of this work was to modify birch wood sanding dust, a waste of plywood production, using amination reaction with epoxy amine for obtaining WPCs. The functionalization of the lignocellulosic waste was carried out by two steps, at first, by its activation with alkaline hydrolysis with the followed amination. The alkaline hydrolysis was implemented with sodium hydroxide solutions. The chemical analysis showed that the main part of the water-soluble products in the hydrolysates were hemicelluloses. The relative content of cellulose in the hydrolysed residue increased from 48.1% to 61.2% with increasing NaOH concentration from 0.15 g/dl to 0.50 g/dl and temperature from 40°C to 90°C. The increased cellulose content, which positively effects on the properties of the reinforcing function of the filler, as well as the lowered hemicelluloses content therein, which positively effects on the capacity of the filler to adsorb moisture, at the highest alkali concentration, indicating that a 0.5% NaOH concentration and 60°C were optimal for the hydrolysis of the plywood dust. The increase of an O/C ratio from 0.98 to 1.16 for the initial and hydrolysed dust, respectively, indicated the growth of the content of phenolic hydroxyl groups in the dust. The amination of the hydrolysed dust particles was carried out with diethylepoxypropylamine. This functionalization allowed to introduce tertiary amino groups, the amount of which in terms of nitrogen varied from 1 to 3%, at the lignocellulosic dust surface. The WPC samples were prepared from recycled polypropylene and the modified dust particles < 100 μm. A twin screw extruder and a molding machine were used for obtaining the composites. The complex functionalisation of the lignocellulosic particles favoured the gain in the mechanical properties of the composites in both tensile and bending strength. Despite the positive effect on the mechanical properties of the composites filled with the microparticles containing the maximal nitrogen content, the excess of the negatively charged groups, which were formed as a side effect during amination, decreased their hydrophobicity as compared with the case of the composite samples filled with the dust particles containing lower nitrogen amounts as indicated by their values of contact angles and water sorption.

SYNTHESIS AND PROPERTIES OF CATIONIC STARCHES OBTAINED BY USING 3-CHLORO-2-HYDROXYPROPYL TRIMETHYLAMMONIUM CHLORIDE

J. Bendoraitiene[1], **K. Almonaityte**[1], **R. Rutkaite**[1], **V. Coma**[2]

[1] *Department of Polymer Chemistry and Technology, Kaunas University of Technology, Lithuania, joana.bendoraitiene@ktu.lt*
[2] *University of Bordeaux, UMR 5629, CNRS, LCPO, France*

Starch is biodegradable, renewable and relatively inexpensive biopolymer, which makes it attractive as an environmentally friendly material for industrial use. Chemically modified starches are generally obtained by treating starch with agents that can react with hydroxyl groups. Such starches have physicochemical properties that differ significantly from those of native starches.

The aim of present investigation was to obtain cationic starches (CS) from native potato starch by using 3-chloro-2-hydroxypropyltrimethylammonium chloride (CHPTAC) as cationization agent with and without the addition of CaO catalyst (Fig.1) as well as to characterize obtained products by FT-IR spectroscopy, X-ray diffraction, SEM, TG, viscosimetric techniques.

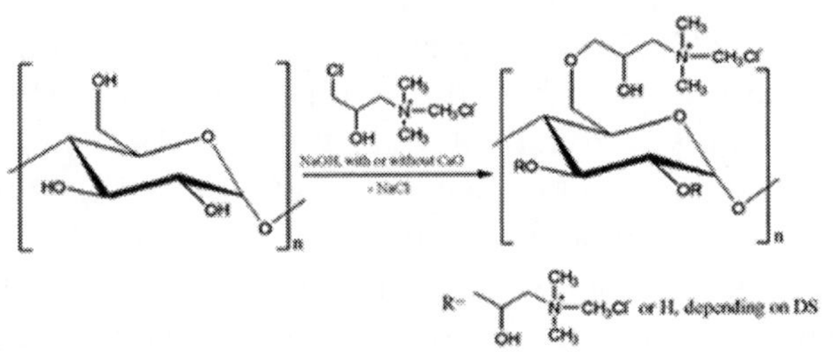

Fig.1. Starch cationization reaction scheme

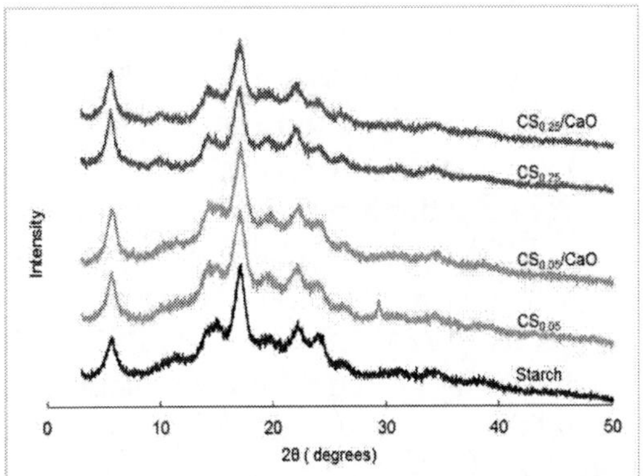

Fig.2. X-ray diffraction patterns of native starch and cationic starches of different DS

X-ray diffraction analysis revealed that degree of crystallinity of synthesized cationic starches was lower than that of native potato starch (Fig.2). FT-IR spectra showed characteristics peaks of native starch and CS. SEM images showed that the size and appearance of CS granules obtained without and by using CaO as catalyst is similar. Thermal degradation of CS of different degree of substitution (DS) occurred at temperatures lower than that of native potato starch. Addition of CaO had no significant effect on CS thermal stability. Gelatinization of modified starches proceeded at lower temperatures than that of starch and peak viscosity of paste during the gelatinization was significantly higher. It was found that by using CaO in starch modification cationic starches of higher DS and better water solubility can be synthesized.

Acknowledgements. The financial support of the Research Council of Lithuania for the Lithuanian-French programme "Gilibert" project No. S-LZ-19-6 is highly acknowledged.

WOOD PLASTIC COMPOSTES ADHESIVE ACTIVITY INVESTIGATIONS TO IMPROVE THE WATER RESISTANCE OF BIRCH PLYWOOD

J. Kajaks[1,*], K. Kalnins[1,2], J. Matvejs[3]

[1] Institute of Polymer Materials, Faculty of Material Science and Applied Chemistry, Riga Technical University, Riga, LV-1048, P.Valdena str.3/7, Latvia, janis.kajaks@rtu.lv*
[2] Troja LTD, Riga, LV-1004, Bauskas str.143, Latvia, [3] Ļatvijas finieris JSC, Riga, LV-1004, Bauskas str.59, Latvia , [2]karlis.kalnins@finieris.lv, [3]juris.matvejs@finieris.lv

Wood plastic composites (WPC) are perspective and widely used material in different branches of industry. One of them is usage of WPCs as coating of plywood. The adhesion problems of the bonding of veneer and plywood are touched in our previous works [1,2]. It was showen that for birch veneer bonding successfully can be used hot melts based on polyethylene and polypropylene. These laminated materials have excellent adhesive strength and water resistance. Specimens endure all the tests which necessary to carry out of plywood [2]. Our investigated composite (industrially produced sheets from PP+45 wt. % PSD) plan to use as the coating of birch plywood but adhesive activity of this adhesive against plywood was not enough. Thereby to improve of the adhesive interaction between the WPC melting component - PP and the plywood surface in the same way as in the bulk of WPC materials the coupling agent MAPP was used (1- 4 wt.%). Evaluation of the bonding quality experiments were done according to the European Standard EN 311:1992. The measurements of bonding strength show positive influence of the MAPP on adhesive strength of the laminated systems PP+40 w.t.% PSD/birch plywood. Bonding strength increase from 0.68 MPa (without MAPP) up to 1.51 MPa (1 wt.%), 2.47 MPa (2 wt.%), 2.74 MPa (3 wt.%) and 2.72 MPa (4 wt.%) of the MAPP respectively. The samples also were prepared with different depth of scratches on protective WPC (PP+45 wt.% PSD). That scratches usually form during of the usage process – shuttering systems for construction industry (average use is 7-15 times). The experimental results of adhesive strength presented in the Table 1 show that average bonding strength between WPC (PP+45 wt.% PSD) overlays and for plywood was higher than between plywood layers.

Table 1. Samples bonding strength after usage.

Damage deep	*Non-used* N/mm^2	*1 time* N/mm^2	*3 times* N/mm^2	*5 times* N/mm^2	*10 times* N/mm^2
0 microns	2.79	2.70	2.77	2.72	2.66
50 microns	2.81	2.66	2.72	2.74	2.80
100 microns	2.68	2.79	2.69	2.67	2.76
150 microns	2.73	2.75	2.68	2.78	2.79
250 microns	2.67	2.80	2.74	2.69	2.71
500 microns	2.72	2.69	2.78	2.66	2.29

In all cases bonding strength between WPC coating and plywood surface is higher than necessary according to the EN 314 Plywood standard (minimum is 1 N/mm^2).

References
1. J. Kajaks, G. Bakradze, et.al. Mechanics of Composite Materials, **45** (6), 643-650 (2009).
2 J. Kajaks, U. Grīnbergs, et.al. Proceedings of Estonian Academy of Sciences, **61**(3), 207-211 (2012).

INVESTIGATION OF END-USE PROPERTIES OF LINEN/SILK FABRICS AFTER DIGITAL PRINTING

I. Tautkutė-Stankuvienė, **L. Simanavičius, E. Kumpikaitė**

Department of Production Engineering, Faculty of Mechanical Engineering and Design, Kaunas University of Technology, Kaunas, Lithuania, indre.tautkute-stankuviene@ktu.edu

Nowadays new textile finishing technologies become very popular in home textile and fashion. One of the new textile finishing methods, which is not enough investigated today, is digital printing. Active and pigment dyestuff are the most popular for this method in textile industry. Technologies as well as final appearance of textile fabric has different final result of these two methods. End-use properties of fabrics after active and pigment digital printing also differ. Thus, the aim of the presentation is to compare appearance and end-use properties of linen / silk woven fabrics after active and pigment digital printing. Also, woven fabrics of new raw material linen / silk were used for the research. It is interesting that two natural fibres of different nature (vegetable and animal) are spinned into one yarn. So, it is the reason for interesting result of the investigation as well.

The pilling resistance of two kinds of digital printing – pigment printing and reactive printing – was analyzed in this study. The diagrams of pilling resistance of both printing kinds were presented in Figure 1.

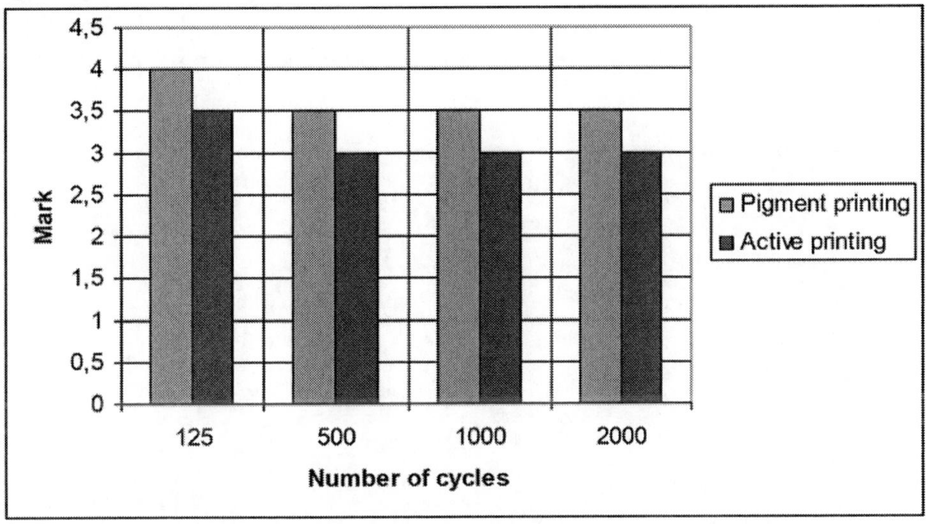

Figure 1. Pilling resistance for linen / silk fabrics after pigment and reactive printing

As it can be seen from the Figure 1, at first, the mark of pilling resistance significantly changes (in 1 mark for pigment printed fabric, and in 1.5 mark for reactive printed fabric) and after that the mark remains the same till the end of the pilling test. It can be this way because of a different method of printing and applying different dyestuff. In pigment printing the dyestuff distributes just on the surface of the fabric and it does not soak into the fabric inside. Before reactive printing, the fabric is soaked and the reactive dyestuff takes up into the fabric. The result of pigment printed fabric was better, because the places with pigment dyestuff on the surface of fabric were more resistant than places, where there was a lack of the pigment dyestuff.

PRODUCTION OF BIODEGRADABLE FIBER MATS BY ROTARY JET-SPINNING

A. Berkovich, P. Ledneva, B. Mankaev, A. Zhirnov, S. Karlov

Chemistry department, Lomonosov Moscow State University, Moscow, Russia, annber@yandex.ru

Biodegradable polymers are promising materials that combine good technological properties and a relatively weak impact on the environment. The growing interest in these materials creates a demand for the adaptation of production and molding methods to the specifics of biodegradable polymers. This paper is devoted to the study of the behavior of some copolymers of polylactide during the molding of nonwoven mats by the method of centrifugal molding. Materials of this type are currently used in medicine as a covering and dressing material, as well as in several other applications.

The method of rotational molding consists in spraying a polymer solution through thin spinnerets of a rotating tank. As a result of the evaporation of the solvent and the extrusion of the resulting jet under the action of centrifugal force, fibers are formed. The method is characterized by simplicity of technological design and relative cheapness, which defines it as being of interest for adapting the molding method to biodegradable polymers.

The laboratory system for rotary jet-spinning was tested on a fiber grade polylactide and several lactide based homo and copolymers. The dependence of spinnability the fibers quality on the molar mass, viscosity of the polymer solution and on the speed of rotation of the forming head, as well as the ability of these fibers to degrade under the action of the enzyme (proteinase-k), is considered.

Fig. 1. Macro and micro images of the poly-L-lactide fiber mats

Acknowledgements. This work is supported by Russia Foundation of Basic Research (project18-29-17029)

MATHEMATICAL EVALUATION OF NANOFIBERS POROSITY

V. Kleivaitė, R. Milašius

Faculty of Mechanical Engineering and Design, Kaunas University of Technology, Studentų str. 56, LT-51424 Kaunas, Lithuania, e-mail: virginija.kleivaite@ktu.edu; rimvydas.milasius@ktu.lt

In order to objectively study the influence of various factors on the structure and porosity of the electrospun nanofibers webs and to compare the results obtained by different authors, an objective method of porosity evaluation is needed. Analysis of literature has shown that the lack of a common method for evaluating the structure of electrospun webs generates uneven results. The method of electrospun web porosity evaluation by maximum pore size in the web is presented in this paper. The main investigations were done with polyamide 6 electrospun nanofibrous web. It was stated that distribution of maximum pore size in various places of the web is close to Gausian normal distribution with some positively skew. This skew is the reason why classical statistical methods cannot be used for electrospun web porosity evaluation. It was stated that for maximum pore size evaluation with reliability 99.9%, it is not possible to use 4S (4 standard deviation) rule. In the case of 100 measurements, it is necessary to use 5S rule, and in the case of 20 or 10 measurements - even 7S or 8S rule, respectively. The results show that for maximum possible pore size evaluation, it is necessary to have much more than 10 or 20 measurements (photos), and for obtaining the same reliability, it is necessary to use a higher coefficient with standard deviation, as in the classical statistical case of Gausian distribution.

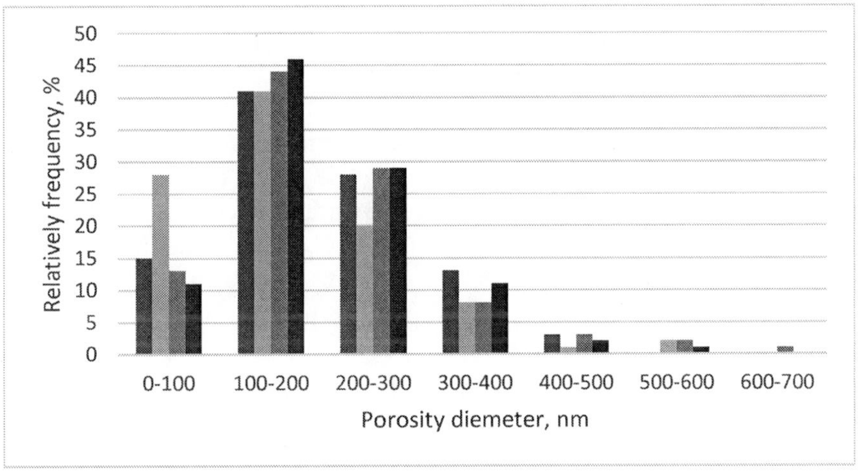

Fig. 1. Distribution of maximum pore diameter.

References

1. J. Malašauskiene, R. Milašius, Fibres & Textiles in Eastern Europe, **18(6)**, 45-48 (2010).
2. V. Kleivaitė, R. Milašius, Autex Research Journal, **18(4)**, 398-404 (2018).

SMART PREPREG AS DAMAGE VISUAL INDICATION SYSTEM

O. Bulderberga, S. Vidinejevs, and A. Aniskevich

Institute for Mechanics of Materials, University of Latvia, Riga, Latvia, olga.bulderberga@lu.lv

In the present work Damage Visual Indication System (DVIS) as prepreg, which includes a fabric impregnated with a mixture of microcapsules with dye, colour activator, and adhesive, is developed. The concept of DVIS is based on the fact that subjected by external loads shells of microcapsules burst, components get in contact and chemical reaction is accompanied by a colour change. Thus, the place of the applied load could be visually identified [1, 2]. As a prepreg such DVIS could be integrated into the composite structure or glued on the surface.
The aim of the work was to evaluate how long the DVIS prepreg keeps the ability to be glued of the surface and evaluate the peel resistance of this system.
The shelf life of the prepreg is an important exploitation property and storage conditions strongly affect on it. During the research, the storing conditions efficiency was determined by the possibility of prepregs to be sticky to an electro-technical textile laminate used as a substrate material and storing the visual damage indication ability (colour changing).
Five samples of DVIS prepreg in a shape of strip with approximately one-meter length and with 2 types of the epoxy binder with or without acetone were prepared. Samples were placed into the refrigerator temperature (+2 °C) and freezer (-18 °C). Starting from the first day little pieces of the prepregs were periodically cut off and destined to gluing to the substrate using soft rubber roller. If the glued piece left fastened on the substrate on the next day, it was thought that the prepreg could continue to be successfully stored. The above check of prepregs was every day for the first 10 days, after three to four days for the next 10 days, and then once a week. The expiration of the storage period was noted when the prepreg ceased to adhere to the substrate.
Experimentally was defined that prepregs with the addition of acetone or without it, as well as both types of binders compositions, keeps ability to be glued to the substrate, for the period around 20 days regardless of storage temperature. By adding acetone ability to be glued to the substrate is prolongated to 43 days when stored at -18 °C, at the same time the amount of added acetone 5 or 20% doesn't affect on it.
To quantify the performance of adhesion or peel ability of DVIS prepregs was used test analogous to Standard Test Method for 90 Degree Peel Resistance of Adhesives (ASTM D 6862). Here two types of fabric (nylon and glass-fiber) were used for DVIS prepregs.
During the peel tests, relation between peel force and time of specimen storage in freezer before gluing was studied. The prepregs with glass-fiber had lower shelf life but had higher peel strength.
The peel tests confirmed results of the previous experiments dealing with the time of storage of the DVIS prepregs with nylon fabric carcasses. In particular, the peel force decrease started after 45 days.

Acknowledgements. This research is funded by the Latvian Council of Science, project DUROCAPS, project No. lzp-2018/1-0084.

References

1. O. Bulderberga, A. Aniskevich, Polymer composite material with inherent function of damage visual indication, ICCM International Conferences on Composite Materials (2017).
2. O. Bulderberga, A. Aniskevich, S. Vidinejevs, A Glass-Fiber-Reinforced Composite with a Damage Indication Function', Mechanics of Composite Materials, **52**, No. 2, 155—162 (2016).

METHODOLOGY OF TESTING OF SMART POLYMER NANOCOMPOSITES PRODUCED BY 3D PRINTING

A. Aniskevich[1], S. Stankevich[1], J. Sevcenko[1], O. Bulderberga[1], and D. Zeleniakiene[2]

[1] *Institute for Mechanics of Materials, University of Latvia, Riga, Latvia, andrey.aniskevich@pmi.lu.lv*
[2] *Faculty of Mechanical Engineering and Design, Department of Mechanical Engineering, Kaunas University of Technology, Kaunas, Lithuania*

3D printing with nano-modified electrically conductive filaments as graphene reinforced polylactic acid (PLA) allows designers to provide their structures additional smart functionality. Structures like apparatus housing, made from this type of composite materials firstly perform load-bearing functions. Secondly these smart structures provide additional functionality as strain, temperature, and damage sensing; active heating and de-icing; improved passive cooling; electrostatic discharge; electromagnetic interference shielding.

3D printed structures have several specific weaknesses and problems: essential anisotropy and heterogeneity of the material; strong dependence of material properties on manufacturing parameters; sensitivity of properties to service and storage environment. Respectively, complex of physical and mechanical properties of the polymer nanocomposites necessary for modelling and optimal design of the smart structures.

Aim of the research was to develop methodology of complex experimental characterization of nanomodified composite materials, produced by 3D printing technology.

Set of basic exploitation properties of 3D printed composite made from neat and modified with graphene PLA filament were measured experimentally. Short-term mechanical properties (elasticity and strength) were determined in tensile tests of dog-bone samples and revealed twofold anisotropy of ultimate characteristics of the composite. Shear properties of the composite were measured in torsion tests of tubular samples. Long term properties were evaluated in creep tests and revealed essential contribution of viscoelastic component on total deformability of the composite both in tension and torsion.

To investigate the electrical properties of conductive composite dual-material, samples were printed with nonconductive and conductive plastic to create objects with embedded 3D conductive tracks. The resistivity of the embedded tracks was measured as 0.15 Ohm·m that essentially differs on the resistivity of filament after extrusion 0.038 Ohm·m. Character of volt-ampere diagrams keep linear that provides possibility for e.g. heating of model structure up to 10-20 K above room temperature with relatively small milliampere DC current.

Developed methodology could be a guideline that helps measuring a set of nanomodified material characteristics necessary for modelling and optimal design of 3D printed structures.

Acknowledgements. Work developed under the M-era.Net research project titled NANO2COM – Advanced Polymer Composites Filled with Novel 2D Nanoparticles. The research was funded by grants No. S-M-ERA.NET-18-1 from the Research Council of Lithuania, No. 1.1.1.5/ERANET/18/02 from Latvian State Education Development Agency.

LAYER RHEOLOGICAL MODEL OF THE FLOW OF SYSTEMS CONTAINING POLYMERS AND SURFACTANTS

E. E. Bibik, E. V. Sivtsov

Saint-Petersburg State Institute of Technology, Saint-Petersburg, Russia, pjeka@yahoo.fr

In rheological studies of various systems, for example, liquid crystal polymers, surfactant solutions, sometimes there is the so-called "shear stress plateau". On rheological curves it can be detected by the constant shear stress with increasing shear rate. This study proposes a rheological model that describes this behavior.

A plastic system is a set of interconnected particles that form a three-dimensional frame with a certain strength (ultimate shear stress) τ_S. The basic property of these systems is thixotropy - the ability for reversible isothermal destruction and restoration of bonds between particles when the stresses in the system change. It can be assumed that the thixotropic system under action of a uniform shear stress can split into several layers with a thickness *s*, parallel to the shear planes (see the Fig.). Here $\gamma=u/L$ is the instrumental shear strain rate of the system, *u* is the velocity of the mutual displacement of adjacent viscometer plates, $L=2s$ is the distance between the viscometer plates. In the Fig. the plot of the fictitious flow rate of a disperse system as a continuous medium is represented by the dotted line A-A, and the corresponding shear rate is the tangent of the inclination angle of the line. Fluid permeating porous formations is also involved in the movement. The local velocity of the fluid *u*(x) inside the layers relative to the rigid frame of a certain layer decreases to zero as the distance x to its middle decreases (solid wavy line in the Fig.). The equilibrium state of the system is quantitatively characterized hydrodynamically (and thixotropically) by the equilibrium thickness of the layers *s*.

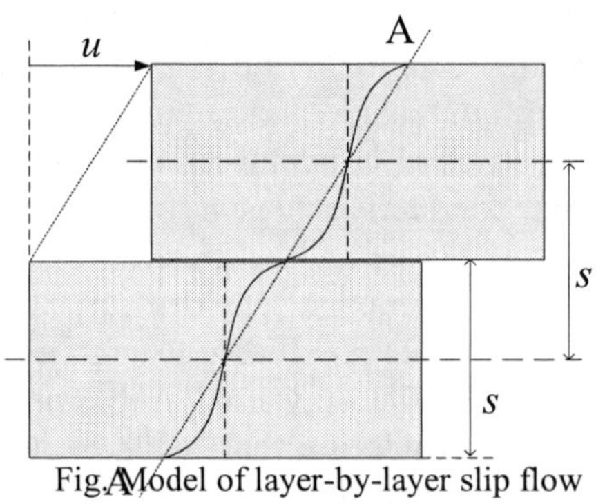

Fig. Model of layer-by-layer slip flow

It was shown that the derivation of the rheological equation of such a system leads to an unexpected result: $\tau(\gamma) = \tau_S$

This means that shear stress in the system does not depend on shear rate and is always equal to the limit shear stress. At low shear rates, when $\tau<\tau_S$, the flow occurs without a change in the structural state: $s=L$, τ increases with shear rate and the viscosity does not depend on shear rate. Another limiting case is the splitting of the layers to a thickness equal to the particle size. Although this state is a theoretical abstraction, in accordance with a known principle a dispersed system behaves like a Newtonian fluid if a further intensification of the flow cannot change the structural state and, consequently, viscosity of the system. Thus, at the layer-by-layer slip of a structured system, all three flow regimes characteristic of the thixotropic systems are realized: flow without disrupting the structure, flow in the state of tixotropic equilibrium value of the structural parameter and flow with complete destruction of the structure.

BLENDING CA WITH PBS TO INCREASE THE BONDING STRENGTH IN TWO-COMPONENT INJECTION MOLDING

M. Klute[1], C. von Boyneburgk[1], H.-P. Heim[1]

[1]*Institute of Material Engineering, Polymer Engineering, University of Kassel, Kassel, Germany,* c.vonboyneburgk@uni-kassel.de

Function integration, elimination of work steps and increased efficiency are only a few of the reasons for processing plastics by the complex process of multi-component injection molding[1]. Integrated production is therefore becoming increasingly important in order to make production processes more sustainable. In this context, two-component injection molding processes, injecting hard and soft components together in a single operation, have become established as an efficient method for producing hard-soft products[2]. At the same time, the pursuit of sustainability and environmental awareness raise customer demand for bio-based polymers[3]. Plastics from renewable raw materials should therefore be made industrially usable for complex parts consisting of several components.

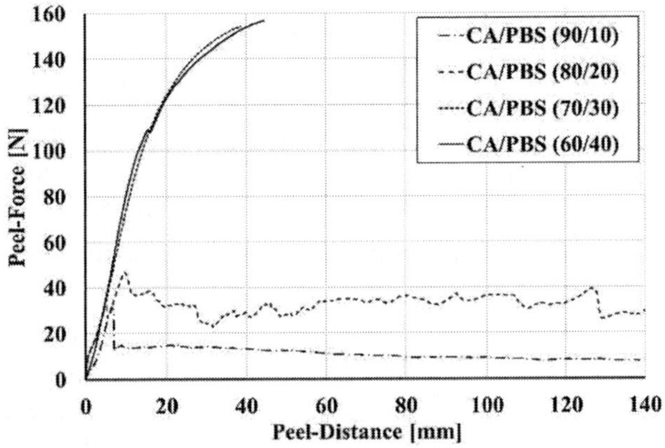

Fig. 1. Peel-Forces measured during the separation of two-component test specimens (hard component: CA/PBS-blends with different weight ratios; soft component: TPU).

Since cellulose acetate (CA) shows no adhesive properties in the two-component injection molding process with bio based thermoplastic polyurethane (TPU), blends of CA and polybutylene succinate (PBS) were produced to reduce the interfacial tension between the materials. While the interfacial tension was calculated from the results of a drop shape analysis, the adhesion strength was measured in peel tests according to the guideline VDI 2019. The comparison of the results gave information about whether the drop shape analysis is a valid method to analyze the adhesive characteristics of material combinations for two-component injection molding. Moreover, tensile tests were performed, to characterize the mechanical properties of the CA/PBS blends. It could be shown, that decreasing the interfacial tension between the two components by blending the CA with the PBS increased the adhesion strength. Adding 30 % PBS caused a cohesive failure of the soft component in peeling tests, showing a bonding strength of at least 147 N.

Acknowledgements. The FNR is kindly thanked for funding this joint project.

References

1. E. Bürkle, Kunststoffe international, **102**, 24-30 (2012)
2. H.-P. Heim et al., Specialized Injection Molding Techniques, Elsevier Science, (2015)
3. M. Koch, Kunststoffe, **103**, 28-34 (2013)

ACTION OF ENZYME ON HIDE STRUCTURE DURING LIME FREE UNHAIRING

V. Valeika, K. Beleška, V. Valeikienė

Department of Polymer Chemistry and Technology, Kaunas University of Technology, Kaunas, Lithuania

Leather industry includes multiple complex processes before final product can be used as raw material for shoes, clothes and other products manufacturing. Chemicals that are used in leather industry (lime, sulphide, chrome compounds) cause almost 80–90 % pollution in the leather industry. The replacement of chemicals by enzymes is an alternative for the reduction of the environmental impact. [1]. The use of enzymes in the industry is gaining more recognition due to advancements made in their isolation and purification, also they ar known as eco-friendly materials [2].

The work analyzes the leather processes and changes of collagen during the processes, when enzymes are used for the lime free unhairing of hide. The unhairing system containing enzyme, sodium sulphide and sodium aluminate was employed for the process.

The investigations show how various unhairing parameters influence hides structure and properties. They were compared to hide properties obtained conventionally. Optimal conditions and the most appropriate amount of enzyme preparation were identified.

It was established that unhairing during 24 hours at 25°C with addition 1% sodium aluminate, 1.2% sodium sulphide and more than 0.1% enzyme preparation leads to qualitative removal of hair but simultaneously to strong effect on collagen. When amounts of mentioned materials are lower than described above, the unhairing quality decreases. The qualitative indexes of the enzymatic process were evaluated and compared with the one after conventional process. The results are presented in Table.

Table1. Indexes of unhairing process.

Index	Unhairing method	
	Experimental (lime free-enzymatic)	Conventional (lime-sulphide)
Unhairing quality	High	High
Amount of removed collagenous proteins during unhairing, g/kg of hide	0.34	0.21
Shrinkage temperature, °C	56.5	56.0
Exhaustion of sulphides, %	70.8	61.3
Porosity of hide, %	67.4	58.4

Pre-tanning processes were tested and obtained after experimental unhairing chromed leather's properties were evaluated. Enzymatic process showed better results in comparison to chemical process, moreover, chromed leathers properties were improved, when such leathers were produced while using enzymatic unhairing process. The results allow conclusion that used enzyme preparation characterizes by good unhairing features.

One the other hand, the IR spectroscopy and thermogravimetric analyses results lead to conclusion that both leathers produced by experimental-enzymatic and conventional-lime-sulphide method are very similar if to compare their structural peculiarities.

References

1. S. Sivasubramanian et al., Chemosphere, **70**, 1015–1024 (2008).
2. J. M. Morera et al., J. Clean. Prod., **112**, 3040–3047, (2016).

CORRELATION BETWEEN RUBBER PARTICLES SIZE AND SOME FUNCTIONAL PROPERTIES OF COMPOSITE MATERIAL BASED ON SCRAP TIRES AND POLYMER BINDER

A. Cirvele[1], L. Malers[2]

[1-2] Institute of Polymer Materials, Faculty of Material Science and Applied Chemistry, Riga Technical University, Latvia, laimonis.malers@rtu.lv

Utilization and recycling of used vehicle tires must be considered as very important way to unload the environment from non-degradable waste [1]. Production of composite materials [2] is one of most perspective ways for the reuse of scrap tires. The produced composite materials can be used in a wide range of applications, for example, in the production industry as an absorbent of mechanical vibrations, in the road industry as protective barriers, in the building industry as sound insulating [3].
In our previous investigations optimization of composition, functional properties and technology of the composite material's production from mechanically grinded scrap tires and polyurethane type binder were realized [4-6].
The special objective of this work is to investigate correlation between selected mechanical properties of the composite material and rubber particles size.
Grinded at room temperature, fractioned rubber crumb (two fractions of rubber particles a and b with size intervals - fine fraction *a* –from 0,5 till 3,0 and coarse fraction *b*- from 2,0 till 10 mm were mixed in different ratio – a/b: 0/100; 25/75; 50/50;75/25; and 100/0) and polyurethane type polymer binder with defined reactivity (C_{-CNO} 8 - 10 %) were used. Composite material samples were prepared by mechanical mixing (at 18-22 °C) of required components ($C_{pol.}$= 8,5 mass %), moulding of samples in specific moulds and hardening of samples under definite conditions (T = 155 °C; 15 min; P = 490 kPa).
Apparent density AD (kg/m^3), Shore C hardness (ISO 7619-1; ISO 868), compressive stress at 10% deformation σ_{10} at static loading mode, tensile strength, elastic modulus and elongation at break (EN 826) in the correlation with rubber particle size were investigated.
The properties and optimization possibilities of composite material consisting of rubber particles with different sizes or fractions and polyurethane type binder were investigated. It was shown that composite material with a ratio 1:1 from fine and coarse fraction mixture of rubber granulate has highest tensile breaking strength, elastic modulus and tensile elongation at break, but the compressive strength at 10 % deformation were higher than expected, while the hardness according to experimental data is relatively low. These differences must be explained by different structure of composite material depending on rubber particles size and their mutual arrangement in material.
hanging the size and quantity of rubber granulate in the composite material.
The obtained results show that variation of composition of the composite material by changing size of rubber granulate have definite influence on the selected mechanical properties of the material. Purposeful selection and mutual combination of rubber particle size included in material can ensure desirable and predictable mechanical properties of composite material.

References

1. J.E. Mark, B. Erman, R. Erich. The Science and Technology of Rubber,3 d. Ed., Elsvier Inc.,USA, 2006.
2. A.H. Hughes, S. Pennington, Precoated rubber crumb for composites, GB Patent 2364708.
3. V.L. Shulman, Tyre Recycling, e.d. Smithers Rapra, Shrewsbury, GBR, 2004.
4. R. Plesuma, L. Malers. Progress in Rubber, Plastics and Recycling Technology, Nr.2, (2015), vol.31, 69-75.
5. R. Plesuma, L. Malers. Open Engineering, Nr.5, (2015), 186-189.
6. R. Plesuma, L. Malers Key Engineering Materials, ISSN:1662-9795, (2016),vol.721, 3-7.

POLYTETRAHYDROFURAN-BASED THERMOPLASTIC POLYURETHANE ELASTOMERS MODIFIED BY FERRITE NANOPARTICLES

A. Bugrov[1,2], G. Vaganov[1], V. Smirnova[1], E. Popova[1]

[1] *Institute of Macromolecular Compounds Russian Academy of Sciences, Saint-Petersburg, Russia, alexander.n.bugrov@gmail.com*
[2] *Saint Petersburg Electrotechnical University "LETI", Saint-Petersburg, Russia*

In the present work, in-situ condensation polymerization method was used to synthesize magnetically active polymer-inorganic composites with the properties of thermoplastic elastomers based on segmented poly(urethane urea)s and $CoFe_2O_4$ nanoparticles. Combinations of symmetric and asymmetric diisocyanates (4,4'-methylene diphenyl diisocyanate (MDI) and 2,4-diisocyanato-1-methylbenzene) with diamines of different lengths (1,3-phenylendiamine and bis[4-(4-aminophenoxy)phenyl] sulfone (BAPS)) were used as monomers in the formation of a hard segment. The soft segment was represented by polytetrahydrofuran (PTHF; M_w = 2000 g/mol) terminated by hydroxyl groups. Cobalt ferrite nanoparticles functionalized by NH_2 groups were used as additional nodes of the polymer network to control the composite supramolecular structure under the action of a magnetic field. All segmented poly(urethane urea)s synthesized in this study had the properties of thermoplastic elastomers according to DMA. However, the most successful combination of the length of soft aliphatic and hard aromatic segments in terms of viscoelastic characteristics was achieved for the composition (MDI-2000PTHF-MDI)BAPS. It can be concluded on the basis of TG curves that neither an increase in the length of aromatic blocks in the structure of segmented poly(urethane urea)s, nor the introduction of nanoparticles as cross-linking agents leads to any significant improve in their heat resistance. It should be noted that both matrixes of multiblock copolymers and magnetic composites based on them exhibit the ability to recrystallize, which allows them to be processed many times over. The incorporation of magnetic filler in the structure of segmented poly(urethane urea)s based on polytetrahydrofuran soft blocks contributes to a significant improvement in the elastic properties of the system. In this case, the surface-functionalized $CoFe_2O_4$ nanoparticles act as plasticizers and make it possible to reach record elongation at break of 1300%, as well as contribute to the extension of the rubber-like elastic plateau from –20 to 270°C.

ANALYSIS OF THERMOPLASTIC ELASTOMER MODIFIED BITUMEN PERFORMANCE OVER A BROAD TEMPERATURE RANGE

T. Ivanova[1], R. Merijs-Meri[1], A. Abele[1], J. Zicans[1], S. Reihmane[1], V. Haritonovs[2]

[1] *Institute of Polymer Materials, Faculty of Materials Science and Applied Chemistry, Riga Technical University, Riga, Latvia, Tatjana.Ivanova@rtu.lv*
[2] *Department of Roads and Bridges, Faculty of Civil Engineering, Riga Technical University, Latvia*

Bitumen has been used as road construction material already since ancient times, just to mention Babylon "Procession Street" as only one example [1]. Since then uses of bitumen in road construction have been vastly extended. According to the recent information of the Eurobitume – a forum for businesses pertaining to the European bitumen industry – the European bitumen industry (EU 28) produces approximately 15 million tons each year from which the majority (85–90 per cent) is used in the construction and maintenance of bituminous or asphalt roads [2].
To cope with increasing stresses of traffic loads, environmental pollution, increased carbon dioxide levels caused global temperature rise there is a necessity for bitumen modification to increase its durability and extend performance range. Among different approaches, bitumen modification with thermoplastic elastomers, particularly, styrene-butadiene-styrene copolymer (SBS), has been demonstrated to be successful in increasing durability, including low-temperature (fracture resistance) and high-temperature (rutting resistance) performance [3]. By considering high price of SBS, constantly various alternatives of polymer modifiers are sought for to improve both low-temperature and high-temperature properties of bituminous materials in spite of the complexity of this multi-factorial problem.
Consequently, the current research is devoted to characterization of performance of polymer modified bitumen compositions over the broad temperature range. Various thermoplastic polymers (e.g. ethylene-octene copolymers with different 1-octene contents) up to the weight content of 8 % have been used to characterize performance of bituminous compositions in comparison to classical SBS modified system. Performance behavior of the developed bituminous compositions have been characterized by using broad range of analytical methods, including fluorescence microscopy, modulated differential scanning calorimetry, rotational rheometry and thermomechanical analysis.
In general, it has been determined that all of the above mentioned methods could be used to characterize certain aspects of performance of bituminous systems over the broad temperature range.

Acknowledgements. The study is co-financed by the European Regional Development Fund (ERDF) measure 1.1.1.1 "Industry-Driven Research" of specific objective 1.1.1 "To increase the research and innovation capacity of scientific institutions of Latvia and their ability to attract external funding by investing in human resources and infrastructure" within the project: No 1.1.1.1/16/A/148 "Innovative use of reclaimed asphalt pavement for sustainable road construction layers".

References

1. J. Connan, Phil.Trans. R. Soc. Lond. B, **354**, 33 (1999).
2. https://www.eurobitume.eu/bitumen/industry/ 01.07.2019.
3. M. Porto et al., Appl. Sci., **9**(4), 742 (2019).

Author Index

Vilnius University
3 Universiteto St.
LT-01513 Vilnius
Lithuania

ISBN 978-1-5108-9526-3

CURRAN ASSOCIATES INC.
proceedings.com

International Conference on Full Scale Ship Performance 2021

Online
10 - 11 February 2021

The Royal Institution of
Naval Architects